FASHIONING TECHNOLOGY

FASHIONING TECHNOLOGY
by Syuzi Pakhchyan

Printed in Italy.

Published by Make:Books, an imprint of Maker Media, a division of O'Reilly Media, Inc., 1005 Gravenstein Highway North, Sebastopol, CA 95472.

O'Reilly books may be purchased for educational, business, or sales promotional use. For more information, contact our corporate/institutional sales department: 800-998-9938 or corporate@oreilly.com.

Print History: July 2008: First Edition

Fashioning Technology Team
Editor: Goli Mohammadi
Designer: Katie Wilson
Copy Editor: Nancy Kotary
Indexer: Patti Schiendelman
Photo Editor: Sam Murphy
Cover and Finished Product Photography: Robyn Twomey
Technical Reviewers:
Matthew Dalton
Kris Magri
Becky Stern

Maker Media Book Division
Publisher: Dale Dougherty
Associate Publisher: Dan Woods
Executive Editor: Brian Jepson
Creative Director: Daniel Carter
Production Manager: Terry Bronson

ISBN-10: 0-596-51437-9
ISBN-13: 978-0-596-51437-2

FASHIONING TECHNOLOGY

A DIY Intro to Smart Crafting

By Syuzi Pakhchyan

TABLE OF CONTENTS

スプーン
2007
年最大の
アルバム
発売!
キーポン!
世界一のロボット

WELCOME

Historically, craft has always embraced technology. Every craft uses a unique set of technologies to shape and manipulate materials. Metalsmithing tools are used to delicately wrap thin strands of gold around semiprecious stones, and sheets of silk are transformed with a sewing machine to provocatively curve around the human body.

Today's digital technologies allow us to print our own textiles with an inkjet printer, knit arm warmers with a knitting machine, and share, discover, and connect through the objects of our own making with online social communities. The old is integrated with the new, resulting in innovative objects blended with rich narratives of the past and the moment. Just as today's craft movement is fueled by current digital technologies and trends, the craft of tomorrow will be equally as influenced and integrated with emerging "smart" technologies of the future. *Fashioning Technology* is a window into the future of craft today.

Fashioning Technology is an introduction to smart crafting, weaving together traditional and unorthodox crafting techniques with new, intelligent materials to create objects and spaces that are interactive and responsive, dynamic and playful. Technology, in the form of digital electronics, is treated just like any other material: it is employed to fulfill both a technical and aesthetic purpose, combining the functional with the symbolic and decorative.

With the availability of new conductive inks, yarns, and textiles, simple circuits can now be silk-screened, hand-sewn, or embroidered, and fashioned into objects of beauty that no longer have to be concealed. The dexterous use of smart materials can transform an old piece of uninspired cardboard into luminescent furniture or a banal strip of industrial felt into haute-tech jewelry.

Fashioning Technology is written for all crafters and makers in the diverse and rich craft landscape who are interested in learning how to use new conductive and smart materials along with electronics to enhance and animate their crafts. My hope is to see felted circuits, quilts embroidered with photochromic threads, posters screen-printed with conductive inks, and much more that I have not yet even begun to imagine. Craft may be deeply rooted in a rich past, but it is also a means of innovation for the future.

—*Syuzi*

1

MATERIALS & TOOLS

// The aesthetic and expressive qualities of every crafted object begin with the choice of materials. This chapter is an introduction to conductive and smart materials, industrial materials, electronic components, and the tools you need to begin crafting with these materials.

New conductive and smart materials present exciting creative possibilities for making objects that are animated, dynamic, and responsive. Smart materials react to their environments — changing shape, shifting color, emitting light, and even producing sounds. New conductive materials, such as inks, threads, and textiles, allow you to easily weave simple electronics into fabrics. By incorporating electronic components and simple circuits into your crafts, you can give your projects a central nervous system capable of "sensing," "expressing," and even "thinking." Even the most basic circuit holds lyrical and emotive potential.

Raw, **industrial materials** from cardboard to industrial felt offer yet another palette to expand your crafting repertoire. A stroll down the aisles of a hardware store can be as inspirational as rummaging through a textile scrap bin. In order to explore this new way of crafting, the traditional crafting workbench is redefined with the addition of a few new **tools**. Your wire cutters will become just as important as your thread snippers as you shift your vintage sewing machine to make room for a new soldering station.

Incorporating **electronics** into your projects and crafting with smart materials is just like learning any other craft: through experimentation and making, you easily become familiar with each material's and each component's unique personality; its technical and physical characteristics as well as its visual and tactile qualities.

NEW CONDUCTIVE AND SMART MATERIALS

Embroidered circuits, pattern-shifting textiles, and hand-sculpted plastic toys are just a few examples of the creative possibilities offered by the seductive world of new conductive and smart materials.

The extraordinary, almost magical, characteristics of these materials lend themselves to fashioning objects and wearables that are artful, quirky, humorous, and poetic. These high-tech materials are a fascinating yet relatively unexplored medium for experimental crafting.

WHAT EXACTLY MAKES THESE MATERIALS SO "SMART"? » Smart or "intelligent" materials are responsive and dynamic. » They have the ability to change color, shape, and size in response to their environments (touch, sunlight, and pressure, for example). » Some even have the capability to remember and return to their original state. » From threads and textiles that shift color to piezoelectric materials that can generate sound with the application of a small current, these materials offer a range of possibilities for crafting objects with an autonomous, secret life of their own.

» Less dynamic but equally as relevant are new conductive materials such as inks, threads, textiles, and even epoxies, which now present an unorthodox fashion to assembling electronic circuitry. Circuits can now be colorful and decorative, embroidered, inked, or knit. They can be exposed, rather than hidden, as a fashion statement or design aesthetic. With the sewing machine as a viable substitute for the soldering iron, this ability to fashion technology enables us to craft a new generation of objects that are interactive, unusual, and fashion-conscious.

Following are concise descriptions, applications, and illustrative examples for a number of new conductive and smart materials to give you a jump start on crafting smart.

Materials Index

Conductive Epoxy

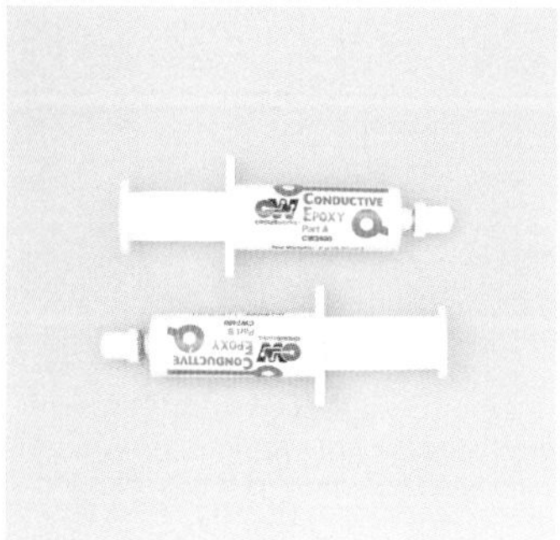

Composition

» Two-part electrically conductive adhesive with copper or silver filaments.

Properties

» An adhesive that allows you to make electrical connections between components and materials where the heat of a soldering iron could be damaging.

Applications

» Traditionally used to repair traces on circuit boards.
» Can also be used to adhere components onto textiles, paper, and other nontraditional materials to build circuitry.

Conductive Fabric and Textiles

Composition

» Textiles plated or woven with metallic elements such as silver, nickel, tin, copper, and/or aluminum.

Properties

» Lightweight, durable, flexible fabrics that have the capability to conduct electricity with low resistivity.
» Can be can sewn like traditional textiles.
» Soft, washable (some), and wearable.

Applications

» Can be used to create flexible and soft circuit boards, pressure- and position-sensing systems, and controls for electronics that can conform to 3D shapes.

Conductive Hook and Loop

Composition

» Nylon hook and loop fastener coated with metallic elements (typically silver).

Properties

» Functions like an ordinary hook and loop fastener (velcro), but has the ability to conduct electricity with low resistivity.
» Can be can sewn like traditional textiles.

Applications

» Makes excellent switches and connection points for electronics sewn onto textiles.

Conductive Thread and Yarn

Composition

» Textile yarn containing metallic elements (typically stainless steel or silver), with nylon or polyester as the usual base fiber.

Properties

» Similar to wires or conductive traces, creates a path for current to flow from one point to another.
» Unlike wires, is flexible and can be sewn, woven, or embroidered onto textiles, allowing for the creation of *soft circuits*.*

Applications

» Offers an alternative method to electrically connect electronic components on a soft and flexible textile medium.
» Enables traditional textile manufacturing techniques, like sewing, weaving, knitting, and embroidering, to be used to make soft circuits.

* Traditional circuits are etched onto copper sheet laminated onto non-conductive substrate; these printed circuit boards (PCBs) are rigid in nature.

Electroluminescent (EL) Ink, Film, and Wire

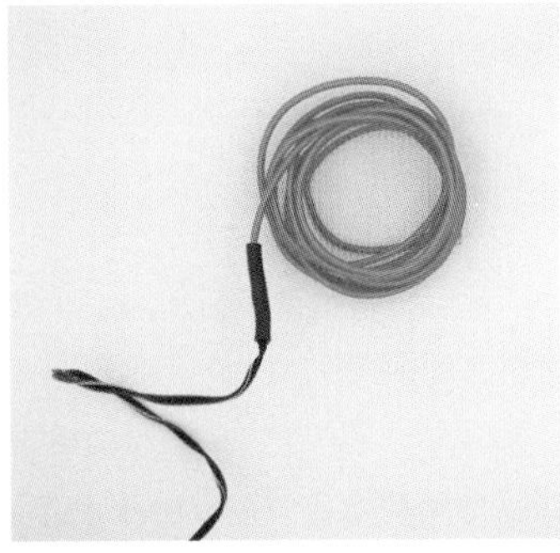

Composition

» Typically comprised of a dielectric layer between two conductive electrodes and a layer screen-printed with phosphor powder.

Properties

» A thin and flexible film or wire coated in phosphor that emits a bright light when electricity is applied, using very little current.
» EL film and wire run on AC voltage. An inverter is typically used to run the film or wire on DC voltage.

Applications

» Typically used for backlighting.
» Film and wire can be used to illuminate curved and 3D surfaces.
» Wire is extremely flexible and can be used to create decorative shapes.
» Inks used to screen-print glowing designs onto polyester or film substrates.

Fiber Optics

Composition

» Plastic or glass cables capable of transmitting light.

Properties

» Transmit light from one end of the cable to the other, essentially acting as a tunnel through which light can travel over a distance.
» Can be bundled together and, because they are lightweight and flexible, send a beam of light from one point to another.
» Transmit light, not electricity, so the light source is isolated from the output.

Applications

» Offer the flexibility to create novel ambient light projects.
» Can be woven into textiles and other heat-sensitive materials.
» Can even be used to bring natural sunlight into indoor environments.

Heat-Shrink Tubing

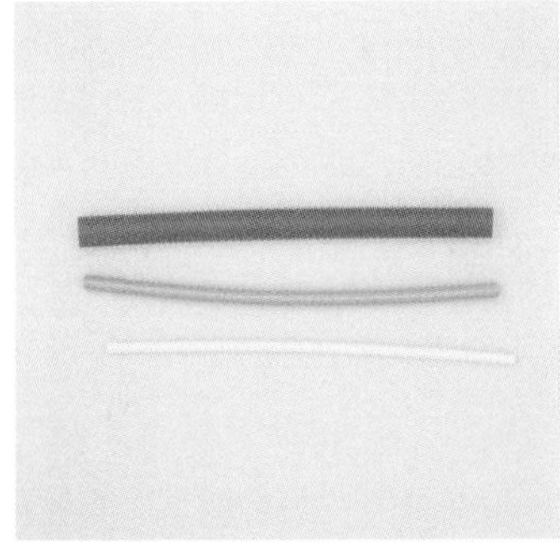

Composition

» Polyolefins.

Properties

» Flexible tube that shrinks when heated.
» Comes in a variety of colors and diameters.

Applications

» Used to insulate raw wires and electronic components.

LEDs (Light-Emitting Diodes)

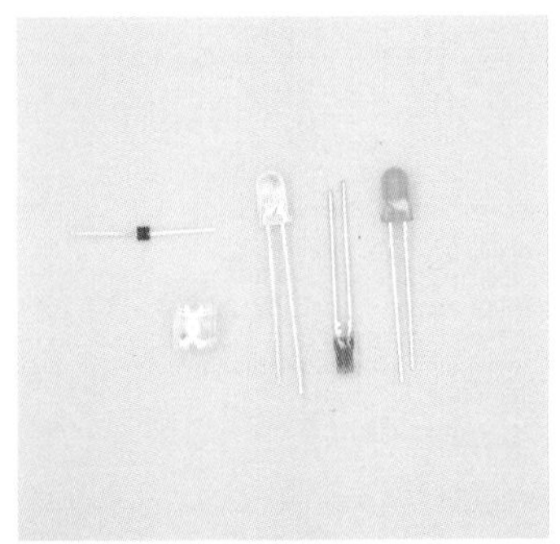

Composition

» Gallium compounds.

Properties

» Small light source capable of emitting bright light.
» Consume very little power, do not emit much heat, and are programmable.
» Come in a variety of colors, shapes, and sizes.

Applications

» Can be incorporated to create programmable, ambient, and decorative lighting.

Magnetic Paint

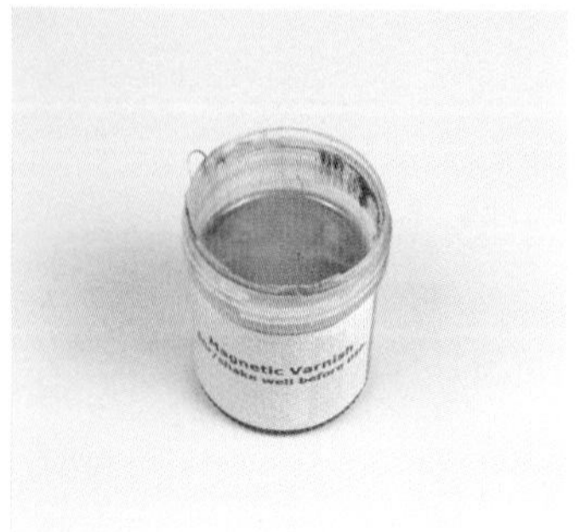

Composition
» Lead-free, water-based latex primer paint mixed with metal particles.

Properties
» Creates a magnetically receptive surface, turning any material (wood, plastic, textiles, etc.) into a surface to which magnets are attracted.

Applications
» Can be used for interior design, converting walls and furniture into magnetic surfaces.

Phosphorescent (Glow-in-the-Dark) Materials

Composition
» Zinc sulfide and magnesium sulfide crystals.

Properties
» Materials such as inks, paints, and thread that emit light over time after they absorb invisible UV light from sunlight or other UV sources.
» Dramatically come to life at night or in low lighting conditions after they have gathered light energy during the day.

Applications
» Can be incorporated to create luminous skins and decorative textures.
» Traditionally seen in glow sticks and glow-in-the dark paint.

Photochromic Inks and Dyes (Ultraviolet)

Composition
» Typically available as powdered crystals comprised of ultraviolet (UV)-sensitive pigments that must be dissolved in the appropriate ink for application.

Properties
» Change from clear to colored when exposed to sunlight, blacklight, or other UV sources. Revert to their original state once removed from the UV source.
» Can change from one solid color to another when mixed with a permanent, colored ink.

Applications
» Can be stenciled, sprayed, and silk-screened onto various media, including paper, plastic, wood, glass, and textiles. For printing purposes, a low mesh screen (between 85–110 threads/inch) is recommended.
» Can be used to create dynamic patterns that change in accordance to lighting changes in their environment.

Piezoelectric Materials

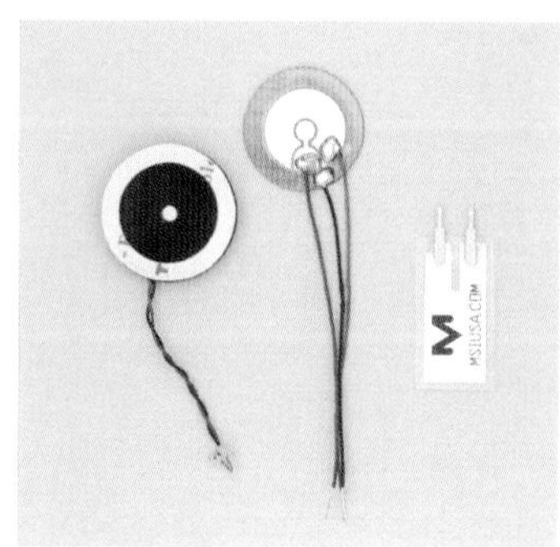

Composition
» Lead zirconate titanate.

Properties
» When subjected to slight mechanical stress (sound, motion, force, or vibration), can generate electrical charges. Inversely, when an electrical charge is applied, these materials can generate a physical force, often enough to be converted into sound. This makes them both sensors and *actuators** at the same time.

Applications
» Serve as excellent environmental sensors that can be used to output and sense sound, motion, and vibration, such as knocking on a surface, for example.
» Coupled with other environmental sensors such as solar cells, can be used to convert light to sound, motion, or vibration, for example.

* An actuator is a device that transforms an electrical input signal into action.

Polymorph Plastic

Composition
» Caprolactone polymer or Oxepanone polymer.

Properties
» A thermoplastic that becomes moldable at around 62° C.
» Becomes pliable and easily hand-sculpted when immersed in hot water or heated using a hair dryer. Returns to a rigid state when cooled.
» Can be reheated and thermoformed indefinitely.
» Available in a variety of colors.

Applications
» A remarkable model-making and prototyping material that can be rolled into flat sheets, sculpted into 3D forms, and used to create molds.

Shape Memory Alloy (SMA or Muscle Wire)

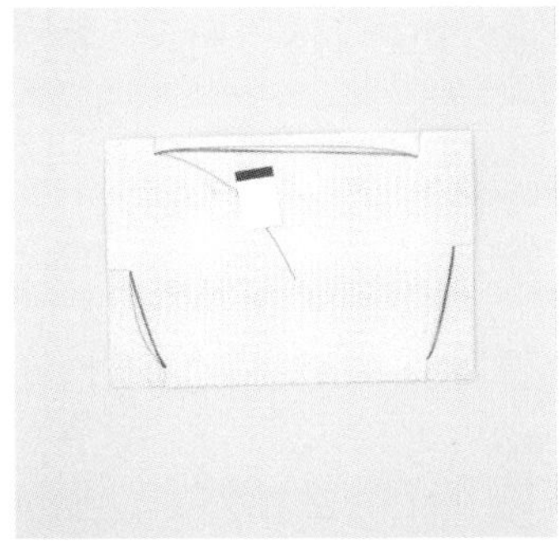

Composition
» A combination of two or more metallic elements, the most popular being Nitinol, composed of nickel and titanium.

Properties
» Unique metals that remember their shape.
» Exhibit hardness and elasticity properties that change radically at distinct temperatures.
» Unlike typical metals, contract when heated and return to their original state when cooled.
» Some SMA wires can be bent into a particular shape by heating at the transition temperature; will then return to their original form when cooled.

Applications
» Can be used to trigger movement.
» Can be woven into textiles.
» Can make fabrics shrink or curl with the application of a small current.
» In robotics, used to animate robots, acting as the robot's muscles.

Solar Cells

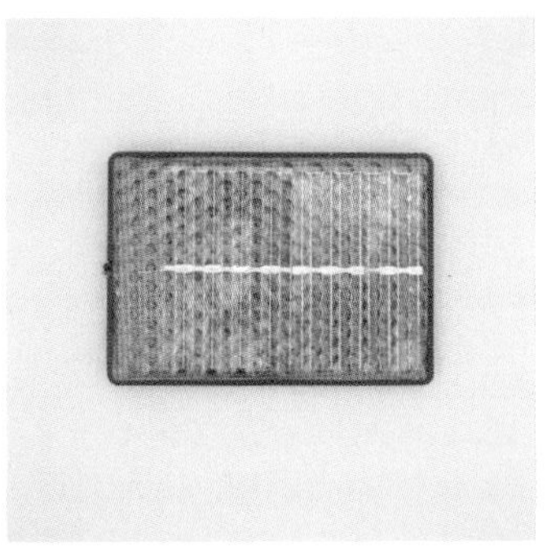

Composition
» Made of a refined, highly purified form of silicon.

Properties
» Convert light energy (typically from the sun) into electrical energy.

Applications
» An excellent sustainable and renewable power source for projects.
» Offer the advantage of acting as light sensors, able to distinguish between light and dark and between different times of day.

Thermochromic Inks (Temperature-Sensitive)

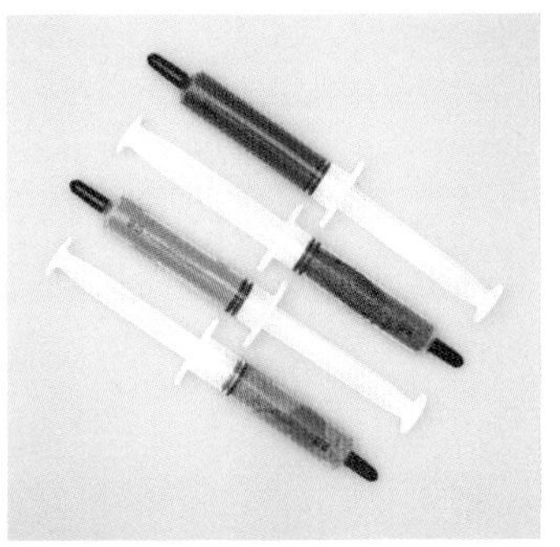

Composition
» Made from various organic and inorganic compounds.
» Pigments must be dissolved in the appropriate ink type for application.

Properties
» Change from one color to another or from color to translucent at a specific temperature.
» Have the ability to infinitely shift color.

» Three main types:
» *low* reacts to cold
» *body* reacts to touch, breath, and body heat
» *high* reacts to hot liquids and air

Applications
» Can be stenciled, sprayed, and silk-screened onto various media, including paper, plastic, wood, glass, and textiles. For printing purposes, a low mesh screen (between 85–110 threads/inch) is recommended.
» Can be used to create dynamic patterns that change in accordance to their environment (for example, to fluctuations in temperature).

INDUSTRIAL MATERIALS

Raw industrial materials — from cardboard to industrial felt — offer yet another palette to expand your crafting repertoire.

Knowing these materials and their capabilities can make a stroll down the aisles of a hardware store as inspirational as rummaging through a textile scrap bin. Here is a list of industrial materials used in a few of the projects.

A. Cotter Pin
A metal fastener with two prongs. Can be repurposed to make flexible battery holders for coin cell batteries.

B. Copper Tape
A thin, conductive foil tape with conductive adhesive. Comes in handy when working with broken solar panels.

C. Flame-Retardant Fabric
Fabric with fire-resistant qualities. An excellent medium for working electronics into textiles for the home. Typically available in wide widths; styles range from velvet and velour to basic canvas.

D. Industrial Felt
A dense, texture-rich fabric made from 100% wool, a renewable and environmentally friendly material that can be used to make virtually anything, from furniture to jewelry. Raw wool felt is typically white or grey, but designer wool felts are available in a variety of colors.

E. Neoprene
A synthetic, rubber fabric commonly used in protective gear and sportswear. Abrasion-resistant, chemical-resistant, waterproof, and elastic, making it an ideal material to craft tech-infused sportswear.

F. Zip Tie (aka Cable Tie)
A fabric or plastic fastener used to bundle loose electric cables and wires together.

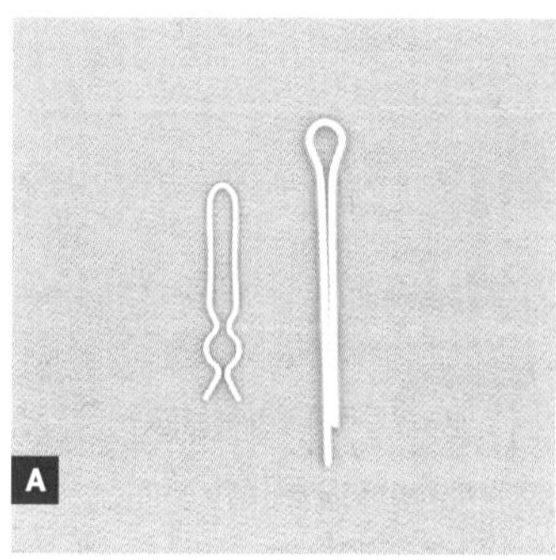

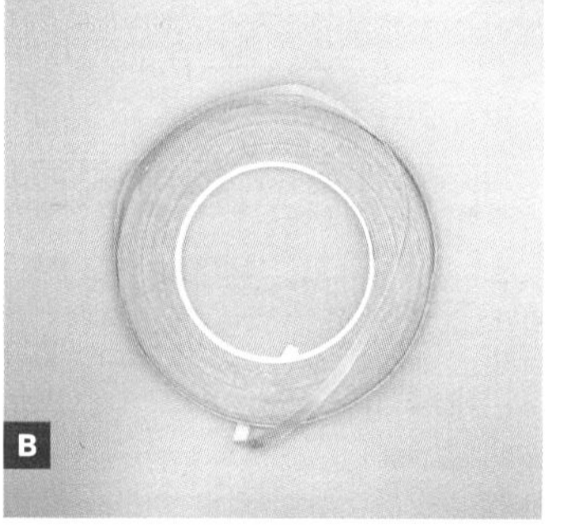

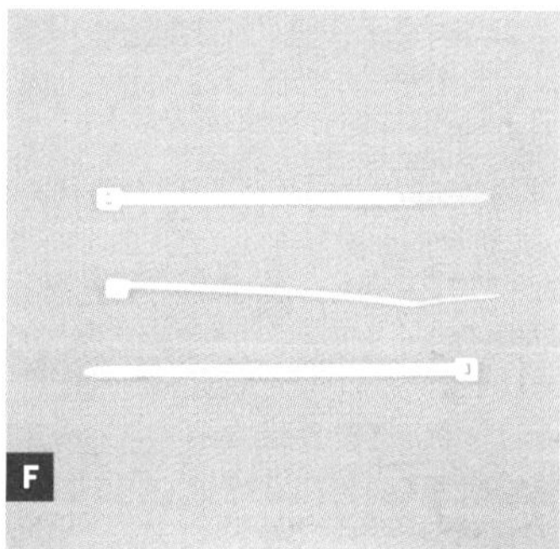

ELECTRONIC COMPONENTS

Electronics are everywhere, from mobile phones and computers to sensing athletic shoes. These sophisticated devices all employ some type of integrated circuit along with a handful of basic electronic components.

» Integrated circuits, or ICs, are the basic building blocks of modern electronics. Bug-like in appearance, ICs are essentially complex circuits etched onto silicon and mounted onto chips. Because all the hard work has already been done for you, they make building sophisticated electronic circuits from scratch a breeze. You will be using ICs such as the Hex-Schmitt inverter in some of the advanced projects and various basic components for the rest.

» Before you start working with ICs or any other electronic components, you need to know what they look like and what they do. Every type of electronic component comes in several variations with different specifications. Even components within the same family may have a dramatically different appearance, and ones that look identical may have very different specifications.

» It is important to pay attention to the *operating values* and the *package type* of each component. Typically, the accompanying specifications and datasheet of each component tell you its minimum, maximum, and normal operating values for voltage and current, and its power ratings. These values should not be exceeded, to prevent damaging your component.

The following pages contain a general overview of the electronic components you will be using to construct your circuits. The more circuits you build, the easier it will become to learn the properties and functions of each.

Components Index

Fixed-Value Resistors

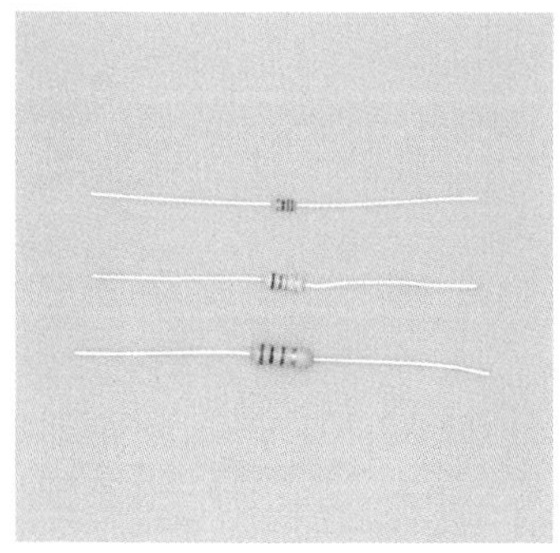

Description
» A cylindrical core with two conductive metal leads, which are not polarized (no negative and positive side).

Function
» To limit the current and divide voltage.

Operating Value
» Two ratings:
» Resistance value rated in ohms (Ω) and a power rating in watts (W).
» Resistors have color-coded bands to help designate their resistance value.

Refer to Chart 1.

Capacitors

Description
» An electrical device with conductive plates separated by an insulating material called a *dielectric*.
» Most common types are:
» mica
» ceramic
» plastic-film
» electrolytic
» *Electrolytic capacitors* are polarized, having one positive and one negative lead, and they resemble little barrels.

Function
» Essentially a temporary battery, capable of storing electrical charge.

Operating Values
» Two ratings:
» A *voltage rating*, which specifies the maximum voltage that can be applied without damaging the component.
» A *capacitance value*, which is rated in farads and is typically printed on the capacitor itself.

Refer to Chart 2.

1 RESISTOR COLOR CODE

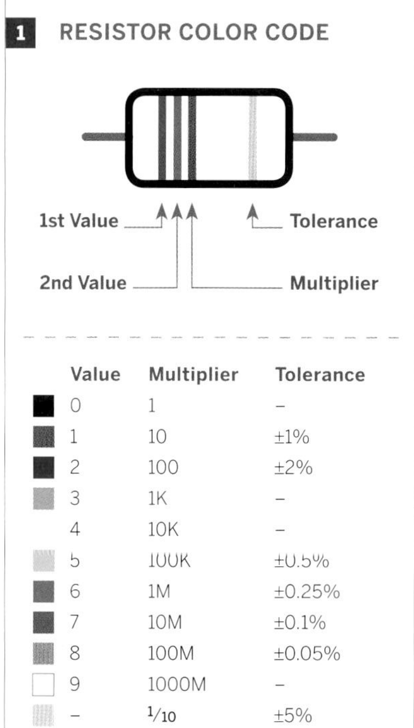

Value	Multiplier	Tolerance
0	1	–
1	10	±1%
2	100	±2%
3	1K	–
4	10K	–
5	100K	±0.5%
6	1M	±0.25%
7	10M	±0.1%
8	100M	±0.05%
9	1000M	–
–	1/10	±5%
–	1/100	±10%
–		±20%

The color of the first band indicates the first digit, and the color of the second band indicates the second digit. The third band indicates the value that the first two digits need to be multiplied by.

IN THE EXAMPLE SHOWN:
The first is 2, the second 6, multiplied by 10:
26×10 = 260Ω resistor.
The fourth band is the tolerance or precision of the resistor.

2 CAPACITOR NUMBER CODE

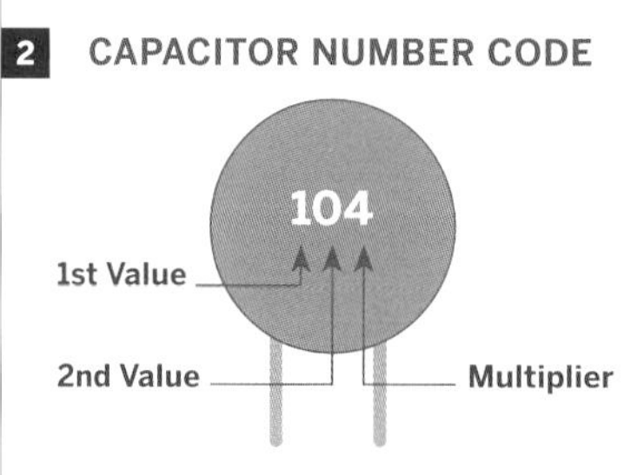

NOTE: For capacitors less than 100pF, only a two-digit number is printed on the capacitor, or a two-digit number followed by a "0". For example, a 55pF capacitor may be marked as "55" or "550".

To determine the value of a capacitor:

The multiplier stands for how many zeros to add to the first two values.

The result is the *capacitance value* in picofarads (pF).

IN THE EXAMPLE SHOWN:
10 with a multiplier of 4 (adding 4 zeros) = 100,000pF or .1μF (microfarads).

SWITCHES

Description

» Used to open (disconnect) and close (connect) circuits by either mechanical or electronic means.

Function

» Switches generally fall into the following six categories:

1. Single-pole single-throw (SPST)
2. Single-pole double-throw (SPDT)
3. Double-pole single-throw (DPST)
4. Double-pole double-throw (DPDT)
5. Push-button (PB)
6. Rotary

» The term *pole* refers to the movable arm in a switch that either opens or closes a circuit. A single-pole switch controls one circuit, while a double-pole switch controls two.

» The term *throw* describes the number of closed positions. A double-throw switch actually has three connections: the right and left close the circuit, and the middle position opens the circuit.

» Switches are also referred to as *normally open* (NO) or *normally closed* (NC). In a normally open switch, the contacts are not touching, so the circuit is open or disconnected. In a normally closed switch, the contacts are touching, so the circuit is closed or connected. If a normally open push-button (PBNO) switch is connected to an LED, when the switch is pressed, the LED turns on, and when it is released, the LED turns off. The reverse is true with a normally closed push-button (PBNC) switch.

A. Magnetic Switch

The magnetic attraction brings the magnets together, opening or closing the circuit. Since magnets are made of alloys, they are conductive.

B. Push-Button (PB)

When pressed momentarily, connects or disconnects contact points. When released, contacts return to original position, as in a doorbell.

C. Reed Switch

Has two thin reeds of magnetic material inside a glass housing. When a magnet is brought near, the reeds magnetize and attract each other, closing the circuit. When the magnet is removed, the reeds separate and move to their original positions. Reed switches can also be normally closed (NC).

D. Tilt Switch

Contains either mercury, a conductive liquid, or a metal ball bearing that, when tilted at a specified angle, connects the contact points, closing the circuit.

E. Toggle Switch

Has a projected lever or arm used to mechanically connect or disconnect contact points, opening or closing a circuit. Can have multiple sets of contact points.

F. Whisker or Trip Switch

Has a thin wire extended out from the component. When lightly touched, triggers circuit to open or close.

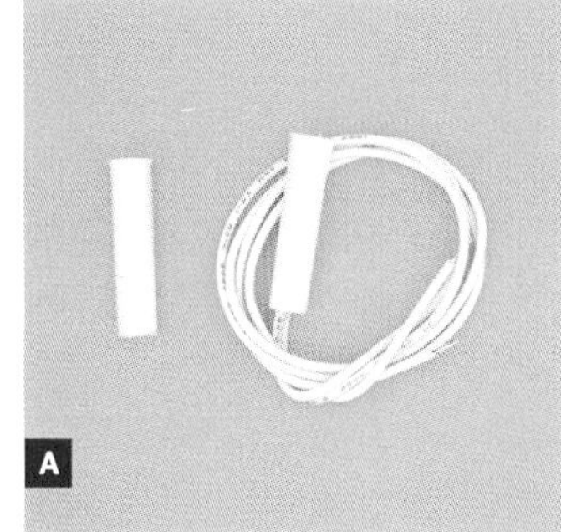
A

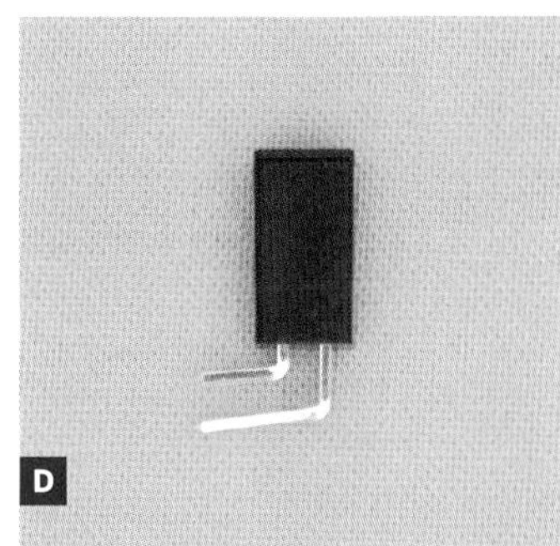
D

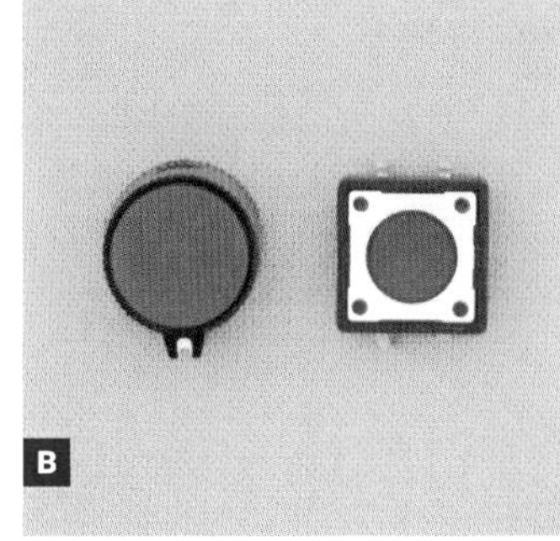
B

E

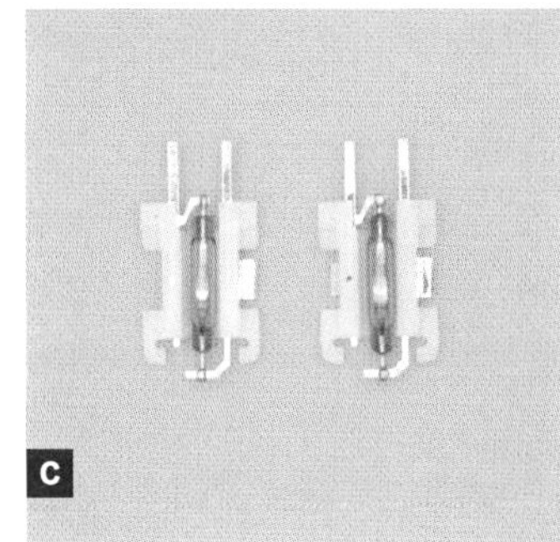
C

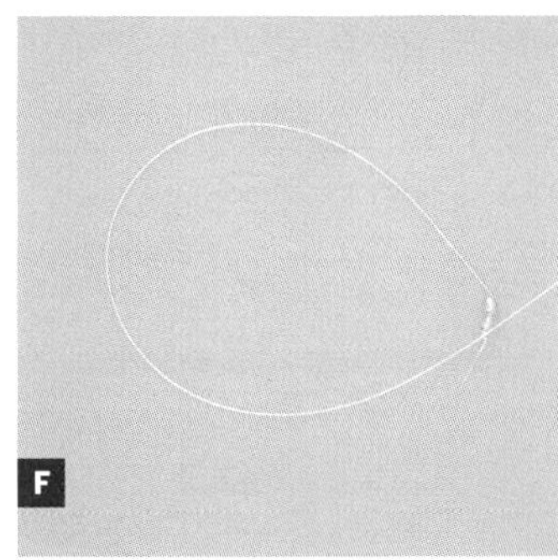
F

Diodes

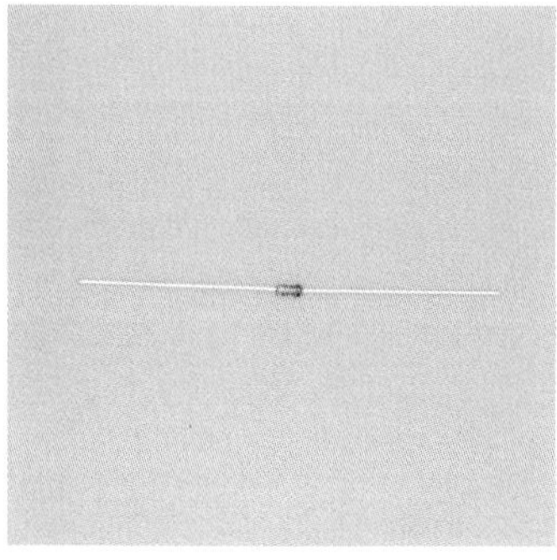

Description

» A semiconductor that allows current to flow in only one direction.
» There are several types that perform a variety of functions. The most common type is a silicon diode. Other diodes include LEDs and photodiodes, which emit light and detect light, respectively. All diodes are polarized with a positive and negative lead.

Function

» Typically used in circuits as a form of protection to maintain a fixed voltage and protect against voltage spikes.
» Also sometimes used as a voltage-sensitive switch.

Operating Values

» Two ratings:
» A *voltage* and *current rating* specifying the maximum voltage and current that can be applied without damaging the component.

Integrated Circuits (aka IC)

Description

» An electronic device made from semiconductor material containing transistors and other electronic components.
» Typically have several pins that need to be connected to other electronic components in a circuit.

Function

» There are countless ICs on the market that perform different functions. You must refer to the product datasheet to get a specific explanation of the application, pin diagram, and operating values.

Operating Values

» Ratings for ICs vary. Refer to the IC datasheet for the minimum, maximum, and typical operating voltage and current values.

Transistors

Description

» A semiconductor device with three *leads*: emitter, base, and collector.

» Functions to allow or restrict current flow, much like a switch, but with electricity as an actuator instead of manual movement.

» Two major types: bipolar and field-effect transistors (FETs). Unlike bipolar transistors, the leads of FETs are referred to as the gate, source, and drain.

» Bipolar transistors fall into two main categories: *NPN* and *PNP*. NPN transistors function similar to a normally open switch, and PNP transistors function as a normally closed switch.

Function

» Commonly used as a current amplifier or an electronic switch.

Operating Values

» Ratings for transistors vary. Refer to the datasheet for the minimum, maximum, and typical operating voltage and current values.

VARIABLE RESISTORS

Description

» Have a predetermined range of resistance value that can be adjusted manually or automatically, unlike fixed-value resistors.

» All variable resistors will have nonpolarized leads.

» Several types available, as detailed here.

Function

» To limit the current and divide voltage.

Operating Values

» Three ratings:

» Maximum resistance rated in ohms (Ω)

» Power rating in watts (W)

» Voltage rating in volts (V)

The power rating defines the maximum amount of current the resistor can handle. The voltage rating defines the maximum amount of voltage the resistor can handle.

A. Flex Sensor

Increases in resistance when bent in one direction.

B. Photocell (aka Light-Dependent Resistor)

Type of photoresistor that varies resistance in response to light levels, typically decreasing resistance as light levels increase. Less common are photocells that increase resistance as light levels increase.

C. Potentiometers (aka Pots)

» Variable resistors with three terminals. The outer two terminals have a fixed resistance between them, from 0 to the maximum predefined resistance. The middle terminal is connected to the wiper, which varies the resistance.

» The most common pot is *rotary*, to which a knob is usually affixed.

» *Trimpots* or *trimmers* are typically used to calibrate the resistance needed in a circuit.

D. Thermistor

» Temperature-sensitive variable resistors that convert temperature change to a change in resistance.

» Two kinds:

» One increases in resistance with an increase in temperature.

» The other decreases in resistance with an increase in temperature.

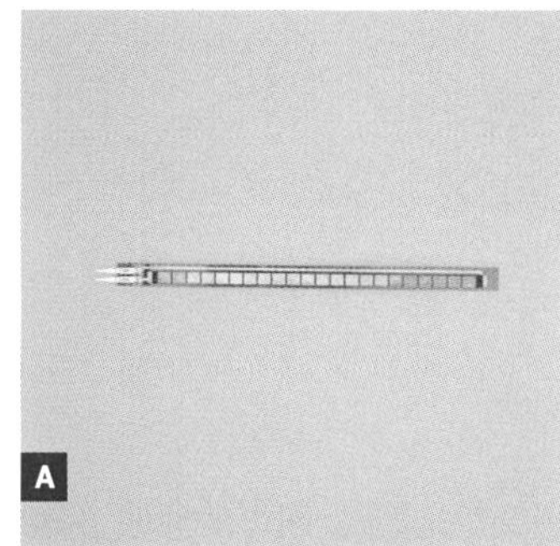
A

C

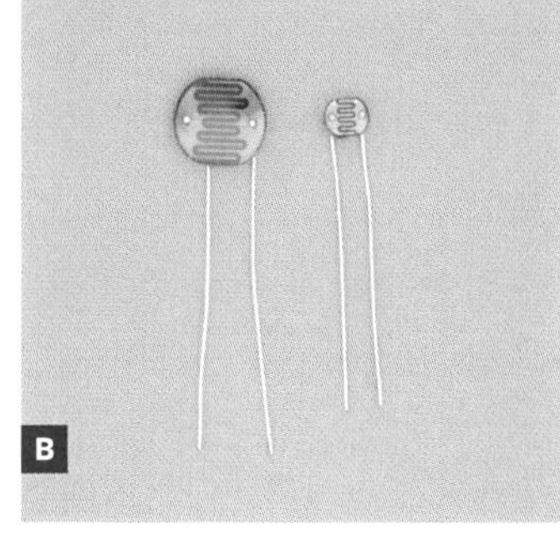
B

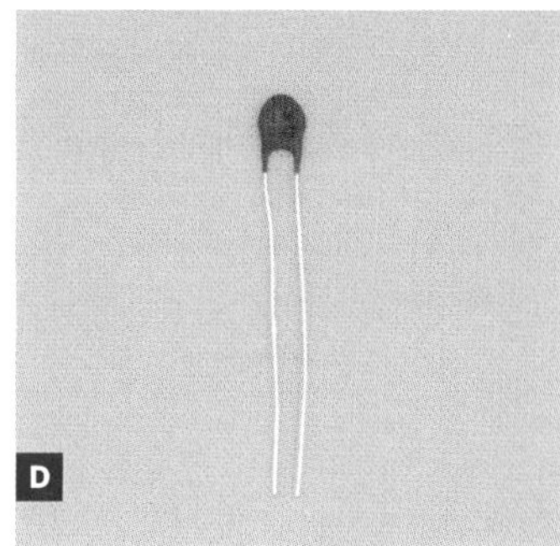
D

YOUR TOOLBOX

Much like any variety of crafting, you need to invest in a small number of tools to get you started. Here is a list of the essential tools to add to your toolbox as you begin working with electronics.

A. Alligator Clips
Provide a temporary electrical connection between electronic components. Each clip has two metal clamps on opposite ends connected by an insulated wire.

B. Electrical Tape
Used to insulate unshielded wires.

C. Third Hand (aka Helping Hand)
An indispensable tool used to hold electronic components and circuit boards in place while you solder. Typically equipped with two small metal clamps to grip components, and a magnifying glass.

D. Multimeter
An essential device that tells you if you have a broken or weak connection, continuity between two components, enough power, and much more.

E. Needlenose Pliers
A pair of small pliers with long, tapering jaws that end at a pointed tip. Used to grip, bend, and curl wire.

F. Perforated Board (aka Perfboard)
Pre-punched board with or without copper traces used to prototype circuits. Unlike a breadboard, electronic components are permanently soldered together.

G. Solder
A metal alloy that is melted (with a soldering iron) to join electronic components. The solder recommended for use in your projects is 60/40 rosin-core solder.

H. Soldering Station
A soldering iron is the fundamental tool used to join electronic components. A typical soldering station comes with a soldering iron, a soldering tip, a stand, and a sponge.

I. Solderless Breadboard
A handy tool used to prototype circuits and temporarily connect all the electronic components together.

J. Wire Cutters
Used to cut wire. Can cut small wires very close up to a flat surface, more so than cutters on pliers.

K. Wire Jumpers
Used to temporarily connect electronic components on a breadboard. They come in different predetermined lengths with both ends stripped and bent at 90°.

L. Wire Strippers
Used to remove the plastic insulation off of wires. Have different-sized grooved teeth for different wire gauges.

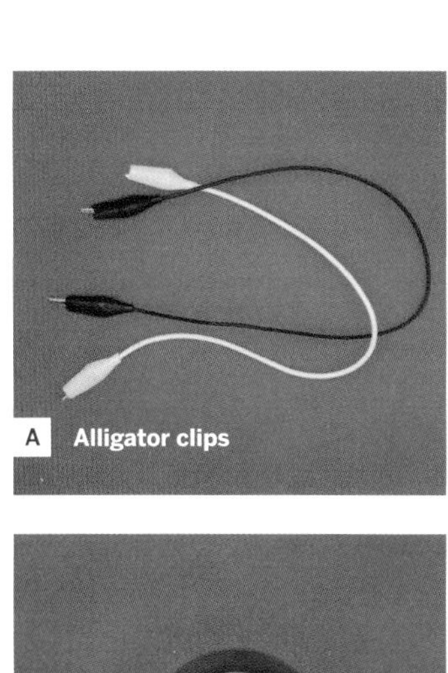
A **Alligator clips**

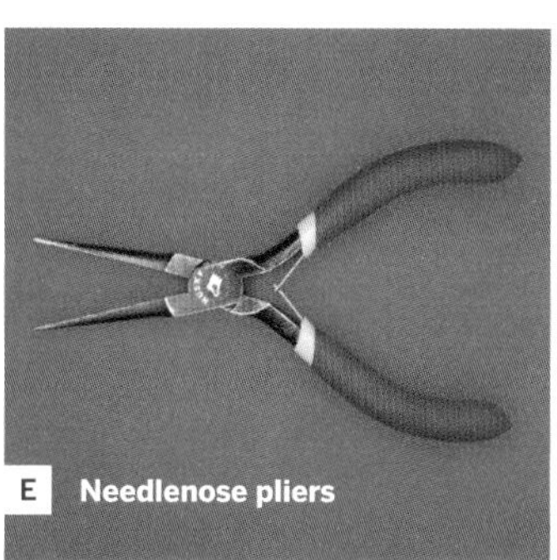
E **Needlenose pliers**

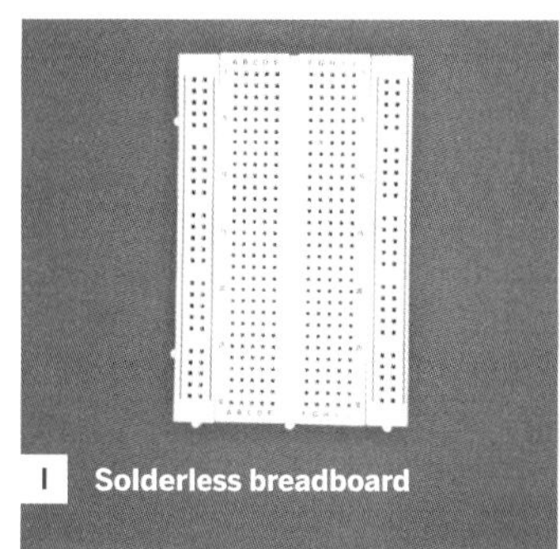
I **Solderless breadboard**

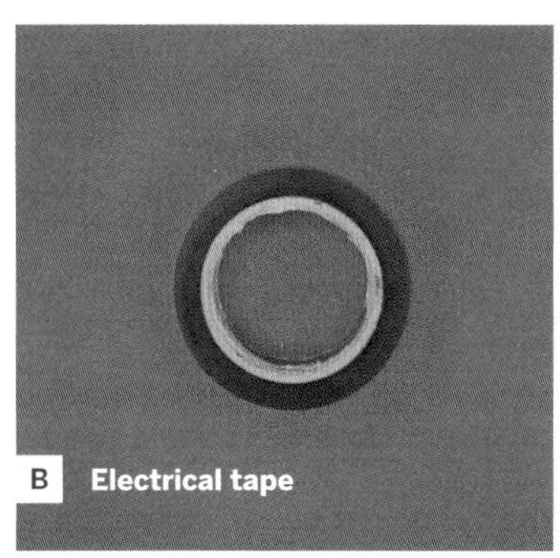
B **Electrical tape**

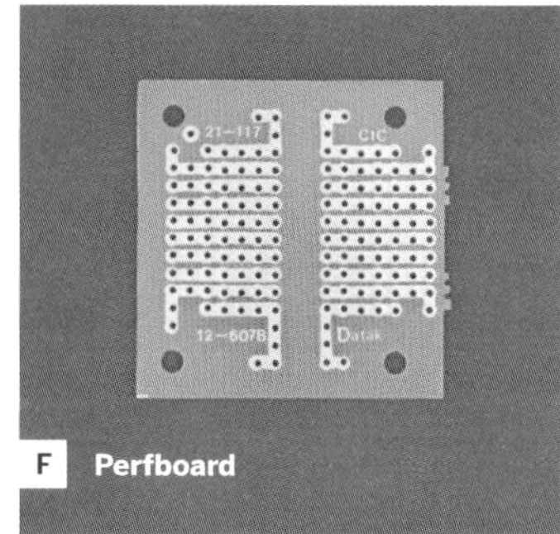
F **Perfboard**

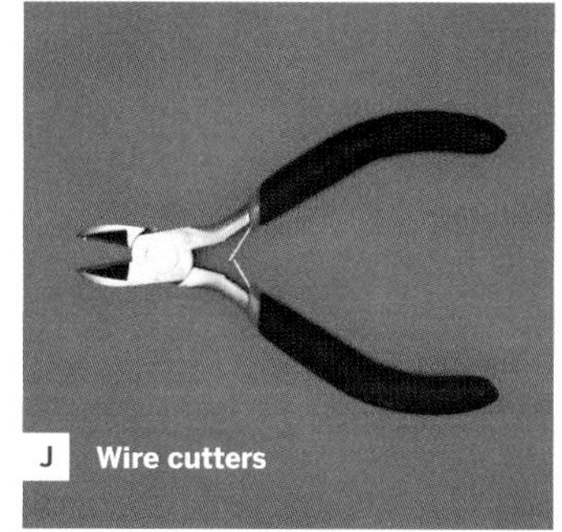
J **Wire cutters**

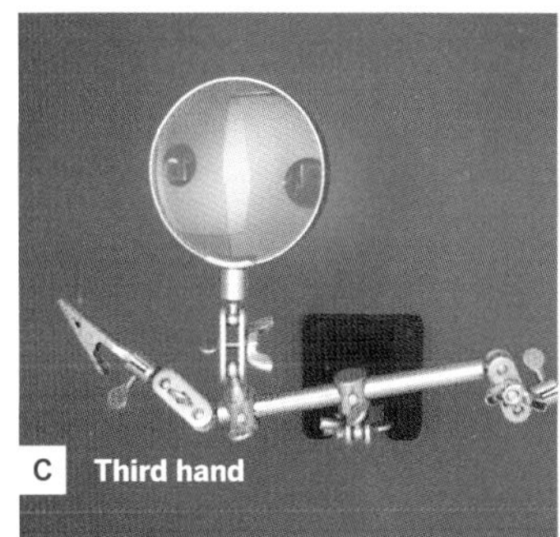
C **Third hand**

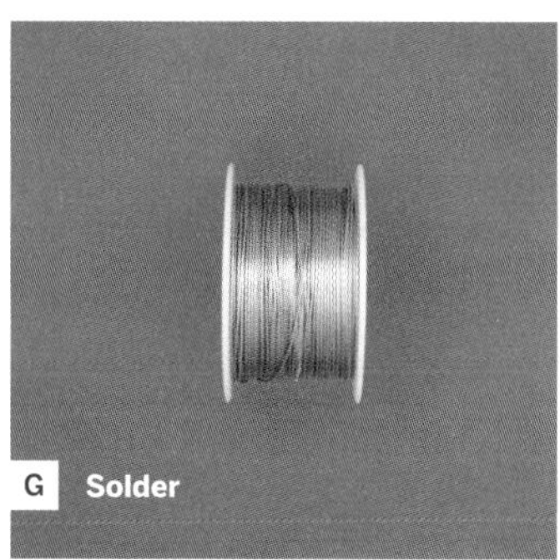
G **Solder**

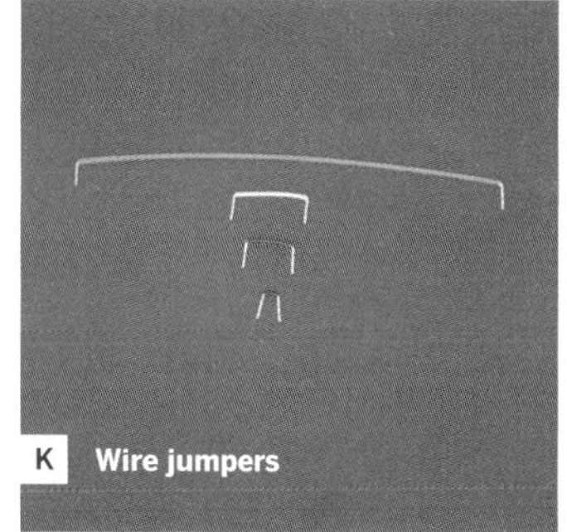
K **Wire jumpers**

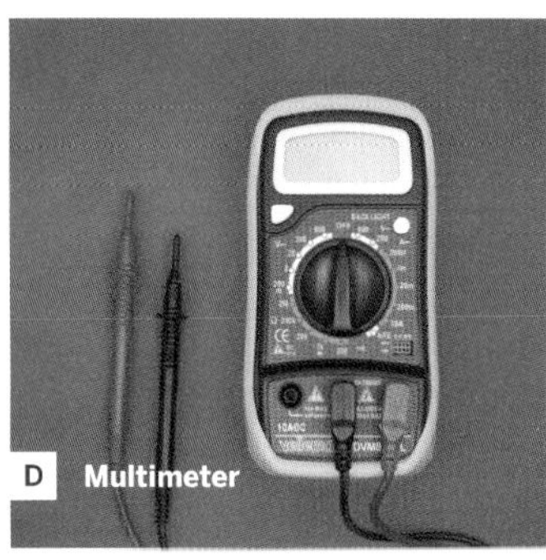
D **Multimeter**

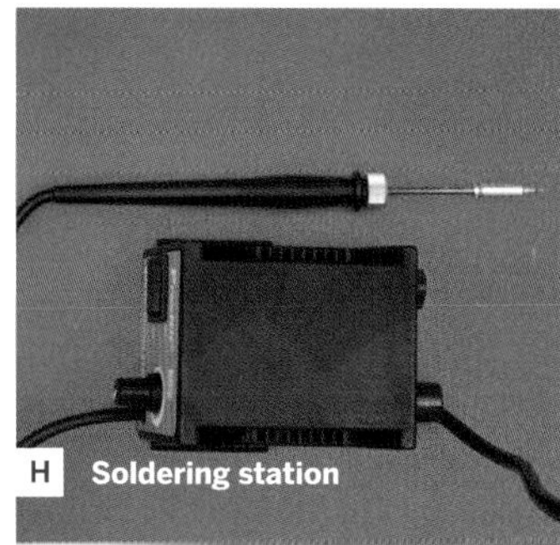
H **Soldering station**

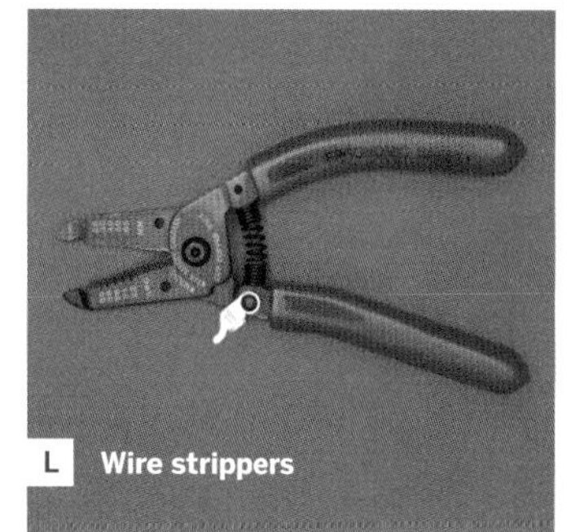
L **Wire strippers**

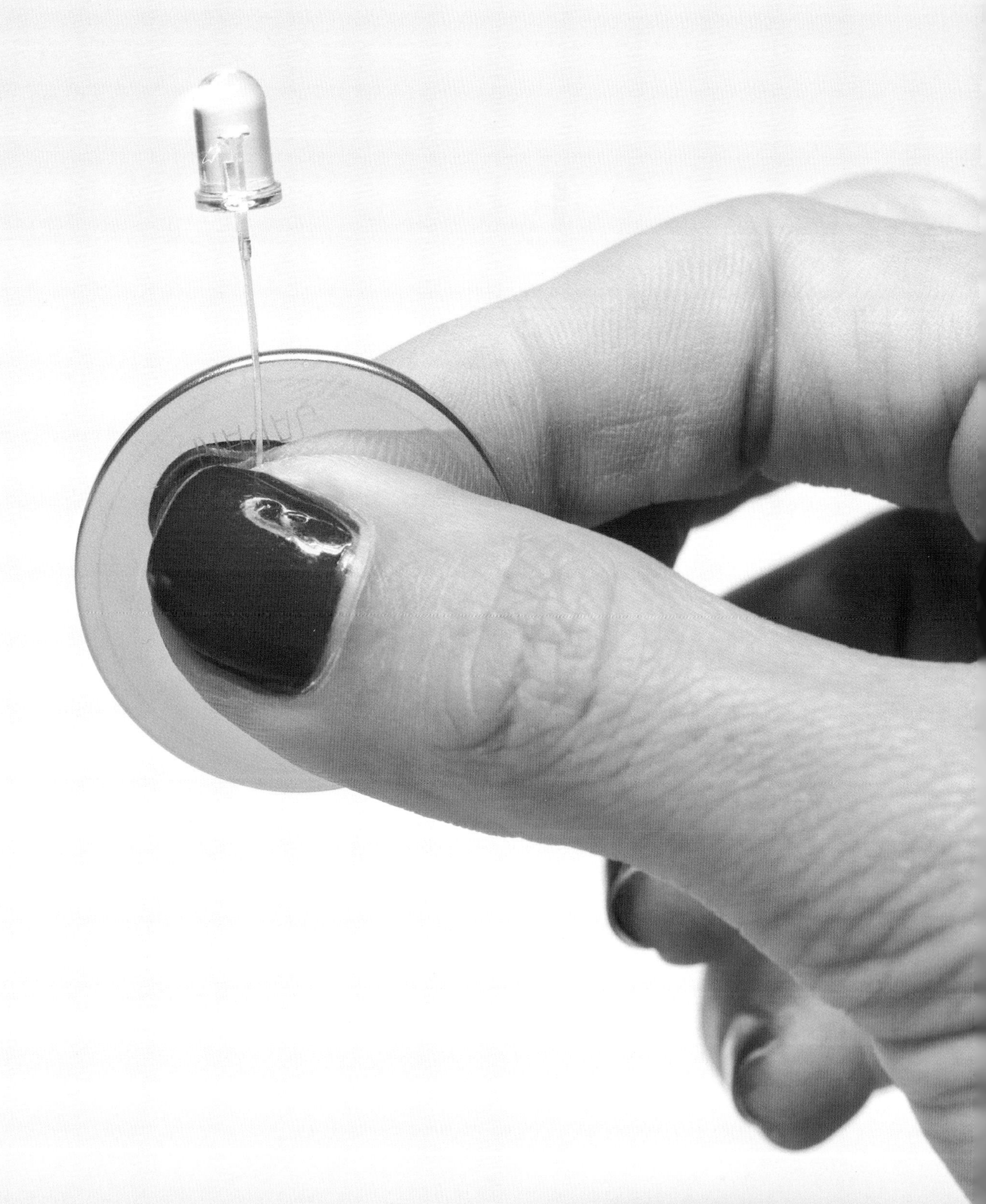

2

// You've mastered the French knot or the perfect blind hem. Soon you will be just as nimble with the soldering iron as you are with the sewing machine.

TECHNICAL PRIMERS

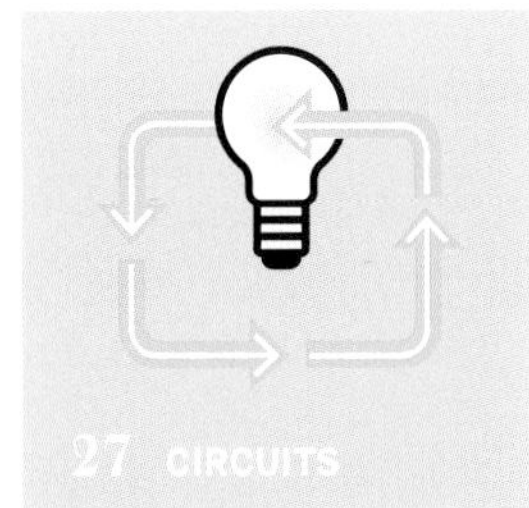

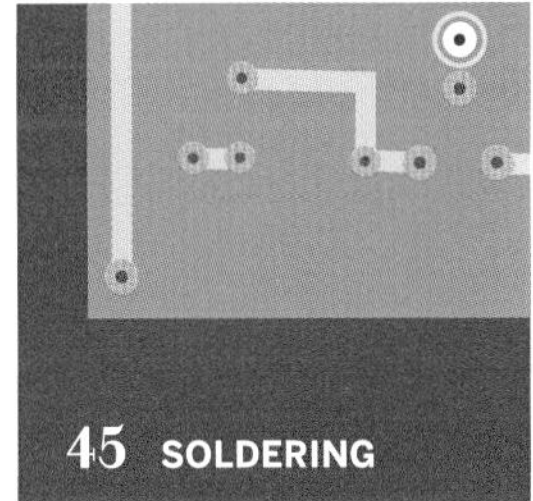

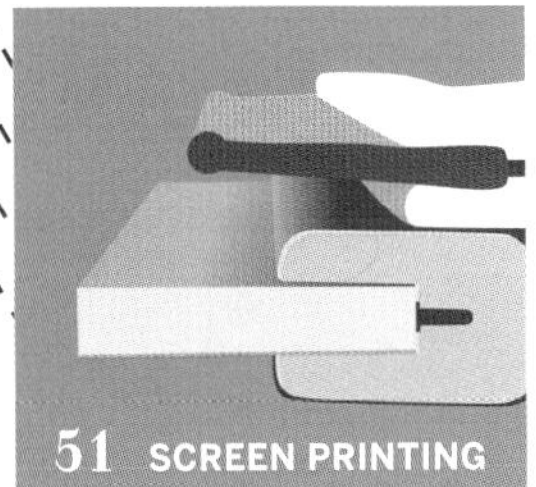

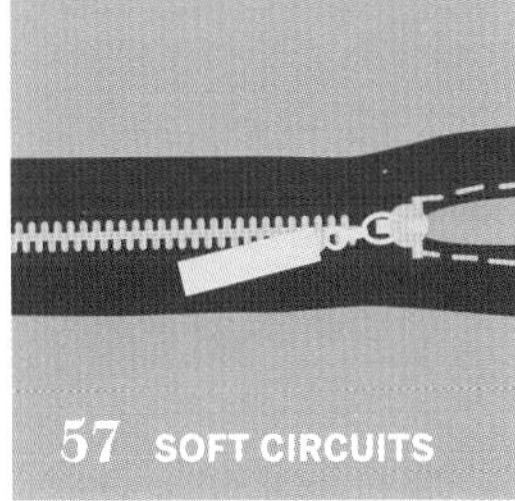

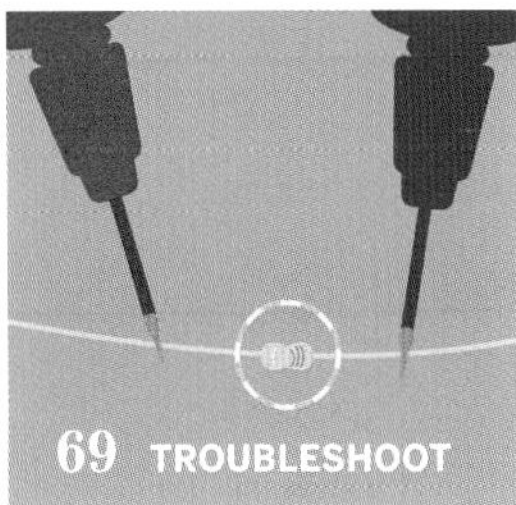

Like any craft, learning the art of electronics takes a little know-how acquired through hours of making, experimentation, and practice. The tutorials and knowledge presented in the following pages are the foundation that all of the projects are built upon. You will learn not only the basics of working with LEDs and how to prototype simple circuits on breadboards, but also how to craft soft switches and screen-print with photo- and thermochromic inks. Equipped with the right tools and fundamental knowledge of electronics, you will quickly begin to integrate technology with your crafts, wiring a chain of LEDs as precisely as you can string a cord of beads.

TUTORIALS

WORKING WITH LEDS

LEDs (light-emitting diodes), the modern craze in lighting, hold the promise of making obsolete Edison's greatest invention, the incandescent light bulb. They are in toys, in automobiles, in stop signs, in every color-shifting gadget, and are even being added to clothing. These techno-sequins are arguably the greatest crafting material invented.

Why are LEDs so darn cool? First of all, they are compact and emit bright light lasting as long as 10 years. Secondly, unlike traditional light bulbs, they don't get hot. And most importantly, they consume very little power. This means that you can continuously run a super-bright LED off of a small battery for more than a hundred hours.

LEDs come in several colors, brightness levels, sizes, and shapes. The two types of LEDs that you will use in these projects are standard LEDs and high-flux LEDs. These and several others are discussed in detail at the end of this section. Before you start working with LEDs, there are a few essentials you need to know.

Powering LEDs and Limiting Current

LEDs are polarized meaning the current from your power source (i.e. battery) can only run through them in one direction.

» The positive side of your battery has to connect to the positive lead (leg) of the LED and the negative side to the negative lead in order for the LED to light up. The positive lead of the LED is referred to as the *anode* and the negative lead as the *cathode*. Notice how your battery has positive and negative sides. If you reverse this order, the LED simply won't light up. Although LEDs are polar, you won't damage them if you plug them in backwards by mistake, so testing polarity is OK.

» In order for LEDs to shine at their best, they need to be powered by the right amount of voltage and current from a battery or other power source. Different-colored LEDs require different amounts of voltage to light up. Red, green, and yellow LEDs typically require between 2.2V and 2.4V. Super-bright white and blue LEDs can require up to 3.4V. When purchasing LEDs, check the specifications on the package. The optimum voltage is usually labeled as the *forward voltage* (V_F).

» Batteries also come in various sizes and voltages. Typically, tiny watch batteries (coin cells) range from 1.5V to 3V. In order to light up an LED, you need to use a battery with the minimum amount of V_F required by the LED, although it is optimal to use a voltage greater than the minimum.

» Most LEDs require 20mAh of current flowing from the battery. Because most batteries supply more than 20mAh, we need to use a resistor to limit the flow of current to each LED so it won't burn out. There are several resistor calculators available online to crunch the numbers for you. By understanding how to read the technical specifications on your LED packaging, calculating the resistor you need is a breeze.

LED CIRCUIT

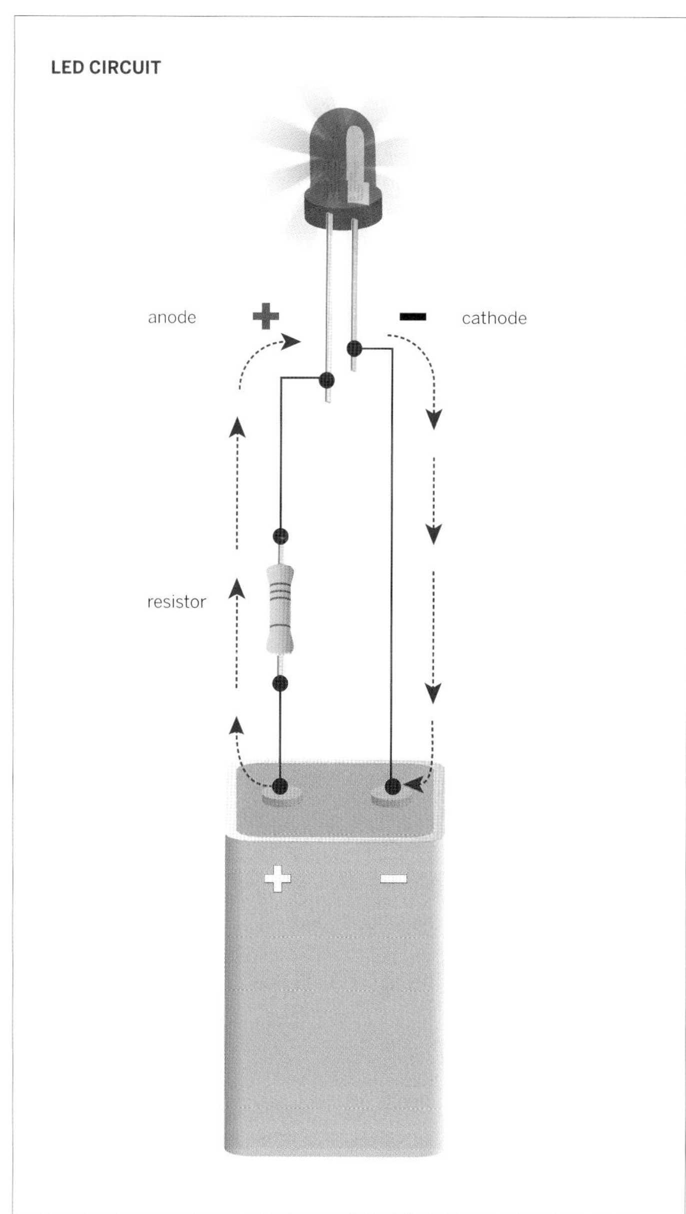

Understanding the Technical Specifications for LEDs

Following is an explanation of all the technical data and symbols that come with your LED package. With an understanding of a few basic terms, you will be able to easily choose the right LED for your project and use online resistor calculators to determine the right resistor for your circuit.

Forward Voltage (V_F): Also referred to as the *forward voltage drop*, the V_F is the minimum amount of voltage needed to light up an LED.

Luminous Intensity (I_V): The I_V is the amount of light emitted from an LED in a particular direction. It is measured in millicandela (mcd). For our purposes, consider luminous intensity the "brightness" of an LED. The greater the millicandelas, the brighter the bulb.

Forward Current (I_F): I_F is the amount of current an LED uses.

Viewing Angle: The viewing angle is the spatial distribution or spread of light. It is expressed in degrees that measure the width of the light beam. LEDs with a small viewing angle produce a more focused beam, and LEDs with larger viewing angles produce a softer, more dispersed beam.

LED SAFETY Light output from high-power infrared (IR) and UV LEDs is intense and may cause eye injury if looked at in close range. UV LEDs are particularly dangerous, as the light emitted is invisible to the human eye. You should never handle UV LEDs without UV-blocking safety glasses.

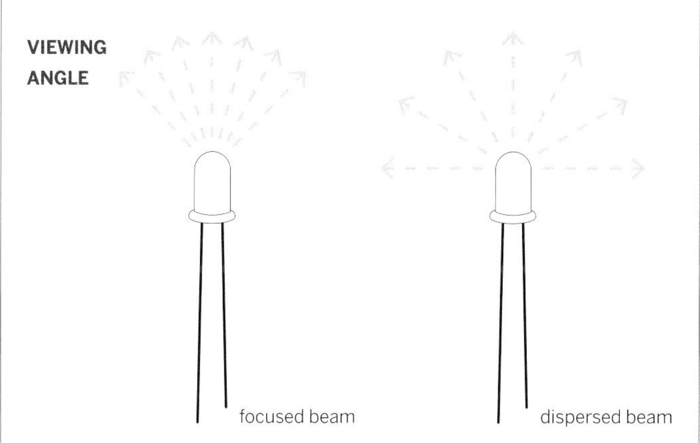

Using Resistor Calculators

There are several resources available online that calculate the appropriate resistor value you should use in your LED circuits. These calculation's are based on Ohm's law.

To use a resistor calculator, you need the following information:

1. Supply voltage, or the voltage from your power source. For example, a typical AA battery provides 1.5V.
2. LED voltage drop or forward voltage (V_F).
3. LED forward current (I_F).
4. Number of LEDs that you want to connect.
5. Whether you are wiring the LEDs in a series or a parallel circuit. The result will be a resistor value in ohms and watts.

Ohm's Law

Ohm's law is the relationship between voltage (E), current (I), and resistance (R) in an electrical circuit. It can be described in the following three mathematical equations:

I = V/R (current formula)
V = I×R (voltage formula)
R = V/I (resistance formula)

When two of the three values in a circuit are known, using the above equations, it is relatively simple to determine the value of the third unknown.

Connecting LEDs

There are three ways to chain several LEDs together to create a string of lights: in parallel, in series, or in a combination of series and parallel.

1. To connect LEDs in parallel, each positive lead is connected to the positive lead of the next LED, and each negative lead to the negative lead of the following LED.

2. To connect LEDs in series, each positive lead of an LED is connected to the negative lead of the next LED, and so forth.

3. Separate LED chains wired together in parallel can be combined all together in series and vice versa.

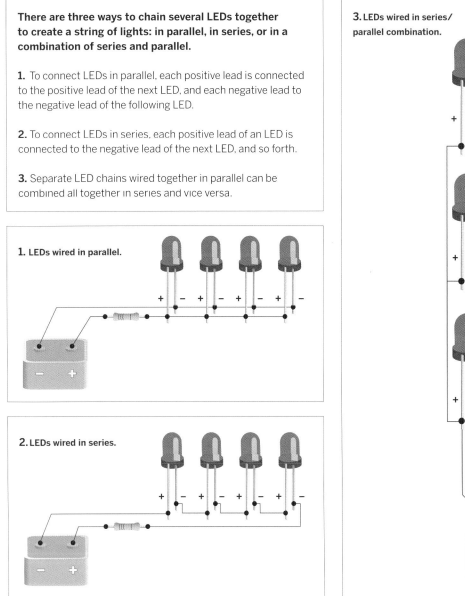

1. LEDs wired in parallel.

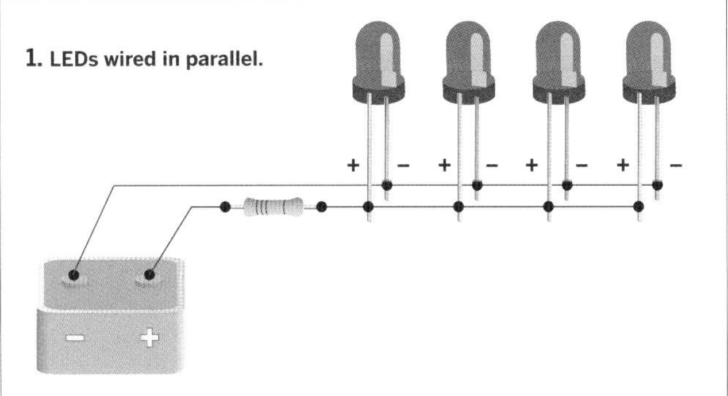

2. LEDs wired in series.

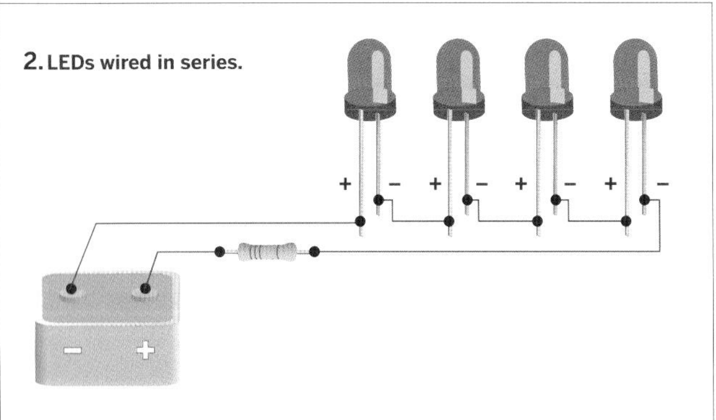

3. LEDs wired in series/ parallel combination.

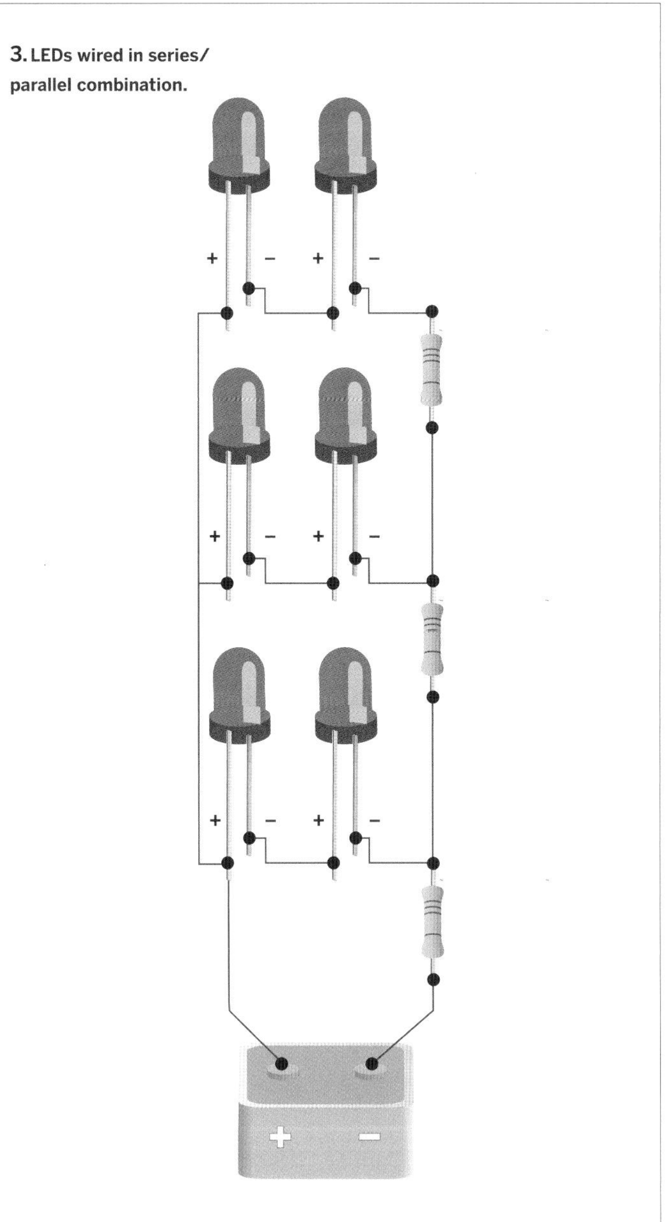

Wiring LEDs in Parallel Versus Series

The decision to wire LEDs in series versus parallel depends mainly on three factors: the power source, the number of LEDs, and whether you're connecting different-colored LEDs together. Here's where a little bit of math comes in.

PARALLEL LEDs wired together in a parallel circuit work great if you need to string together a number of same-color LEDs and have them powered by a small voltage source.

Electricity likes to take the path of least resistance. If you have two different-colored LEDs, one red with a V_F of 2V and the other green with a V_F of 2.4V, the current will flow through and light only the red LED. The green LED will be circumvented completely. It is possible to mix colored LEDs with the exact same specifications (different colors requiring V_F of 2V, for example) in a parallel circuit, but often some of the LEDs end up appearing slightly dimmer than the others.

HERE'S WHY When LEDs are connected in parallel, the voltage across each LED remains the same, and the current is divided between them.

FOR EXAMPLE A circuit with a 3V battery and its appropriate resistor will light up two red LEDs (with a V_F of 2V) wired in parallel. In this circuit, each LED will receive its required 2V, and the current will be dispersed between the two LEDs — therefore, you will drain your battery more quickly than if you had wired your LEDs in series. **Why not wire your LEDs in series, then?**

SERIES When LEDs are connected in series, the voltage is divided equally across each LED, and the current remains the same. In the previous (parallel) circuit, were the LEDs wired in series, they simply would not light up, as they would each receive 1.5V rather than 2V.

So how do you connect different-colored LEDs together? Easily. The power supply must be equal to or more than the sum of the voltage requirements for each LED accompanied by the appropriate resistor. If you want to mix colors and wire your LEDs in parallel, then you need to add the appropriate resistor to each LED (rather than one for the entire circuit), leveling the amount of power required for each to light up.

IN SUM **If your project requires the use of multiple LEDs and a small power supply such as a battery, parallel wiring is the way to go. If your project doesn't preclude such conditions, then wiring your LEDs in series will provide a more stable operation.**

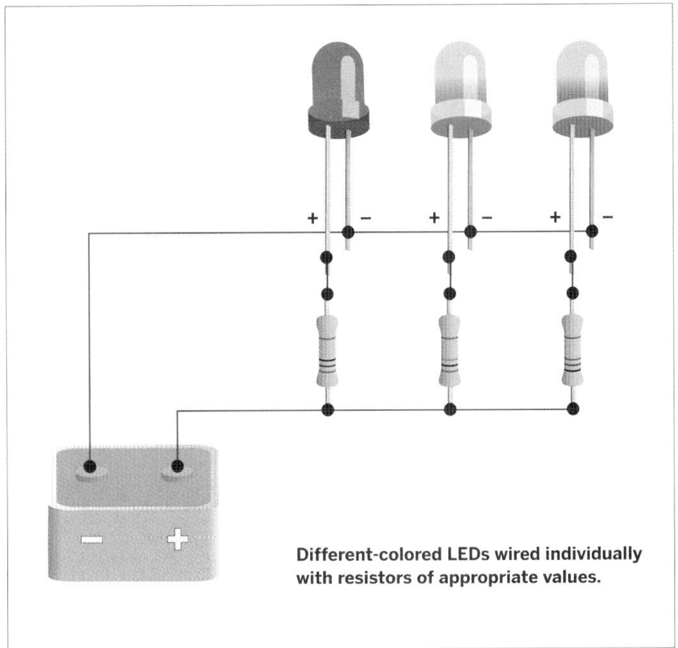

Different-colored LEDs wired individually with resistors of appropriate values.

Different Types of LEDs

There are a plethora of LEDs on the market, ranging in color, shape, brightness, and viewing angle. This is an overview of the different types of LEDs and their characteristics to help you determine the most appropriate type for your projects.

Bicolor or Tricolor (RGB) LEDs

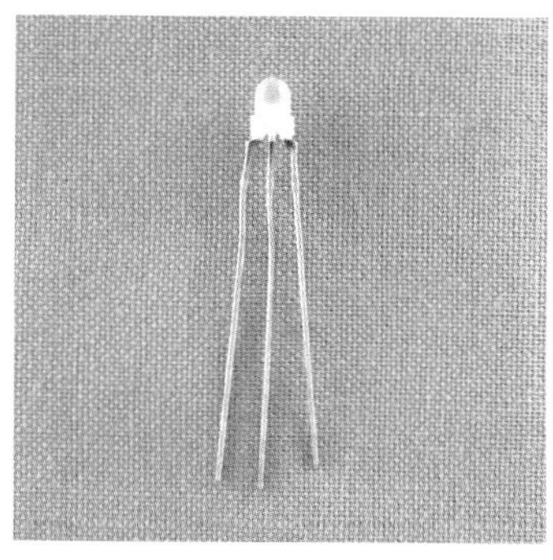

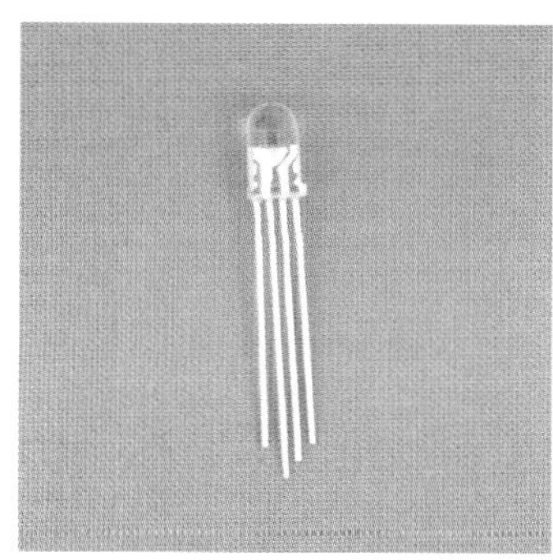

» Two- or three-colored LEDs sandwiched into one housing.

» Bicolor LEDs have three different-length leads, sharing either a positive (anode) lead or a negative (cathode) lead.

» Refer to the data sheet to distinguish the leads, or create a simple circuit using a pair of alligator clips, a battery, and a resistor (see page 29).

» Typically, in bicolor LEDs, only one of the colored LEDs can be lit at a time.

» Tricolor (RGB) LEDs combine red, green, and blue color spectrums, and when mixed together can reproduce a wide range of colors.

» RGB LEDs have four different-length leads, sharing either a positive lead or a negative.

» Typically, a microcontroller is used to program an RGB LED to gradually shift from one color to the next.

» Rare but available from a small number of retailers are RGB flashing LEDs that are preprogrammed to blink and fade from one color to the next.

Blinking/Flashing LEDs

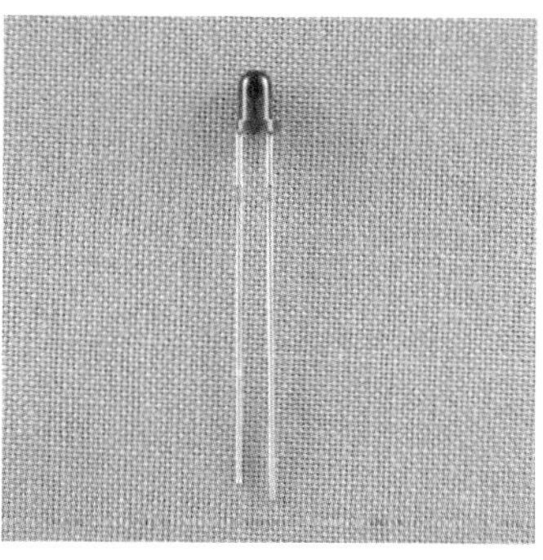

» Have similar characteristics to standard LEDs, except that they contain an integrated circuit (IC). This IC blinks the LED (turns it on and off) intermittently at predetermined time intervals.

» Designed to be directly connected to a power supply, eliminating the need for a resistor.

Infrared (IR) LEDs

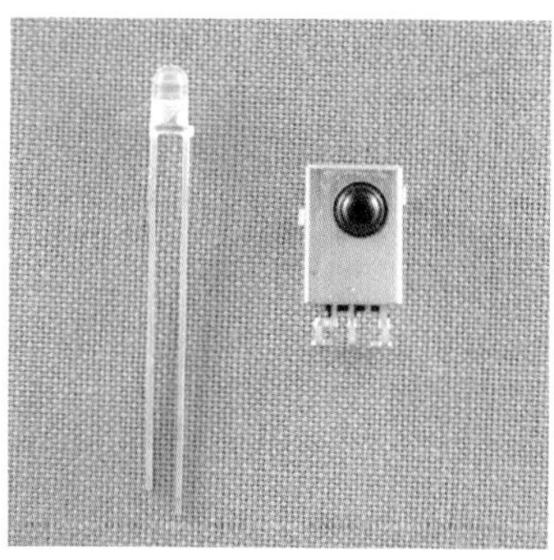

» Emit infrared light that is invisible to the human eye.

» Typically coupled with an IR receiver/sensor (aka detector) that recognizes the infrared light and translates it into an electrical signal.

» The most common example is found in your remote control.

Piranha or High-Flux LEDs

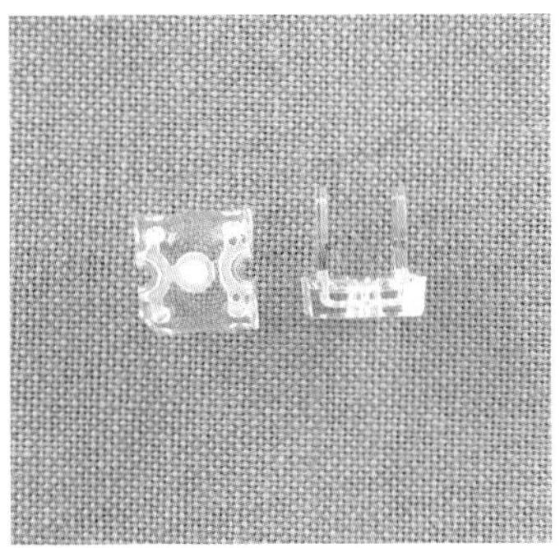

» Square with four leads: two positive and two negative. As all four leads are the same length, to distinguish the positive from the negative you must either refer to the LED's data sheet (usually available as a PDF download online) or create a simple circuit using a pair of alligator clips, a battery, and a resistor (see page 29).

» Available in a number of colors, brightness levels, viewing angles, and a few select sizes.

» Practically all come in a clear case and are brighter than typical LEDs.

» Great for sewing applications, as they can be positioned flush to fabric and their leads are pliable — able to be bent easily and sewn.

Standard LEDs

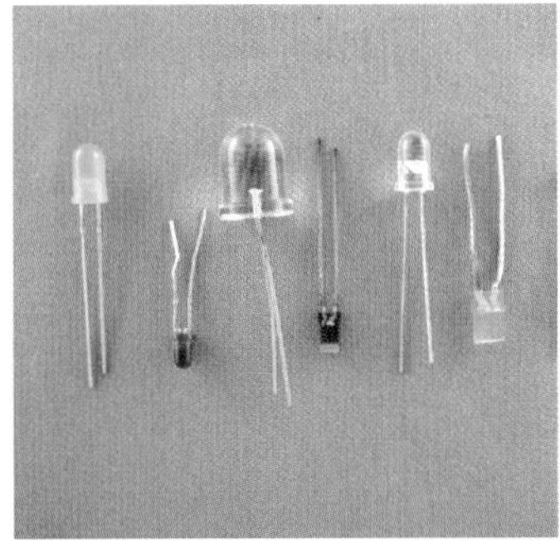

» Have two leads: a positive and a negative. Typically, the long lead is the positive (the *anode*) and the short the negative (the *cathode*).

» Available in different shapes, sizes, and colors. Normally range in size from 3mm to 10mm and come in red, orange, amber, yellow, green, turquoise, blue, and white. Pink and violet LEDs are rare but also available.

» Come in a colored, water-clear, or diffused epoxy resin case. Either round, flat, or square in shape.

» Apart from their physical properties, they are available in different levels of brightness, measured in millicandela (mcd), and in different viewing angles.

» The brighter the LED and the greater the viewing angle, the more expensive the LED.

Surface Mount Device (SMD) LEDs

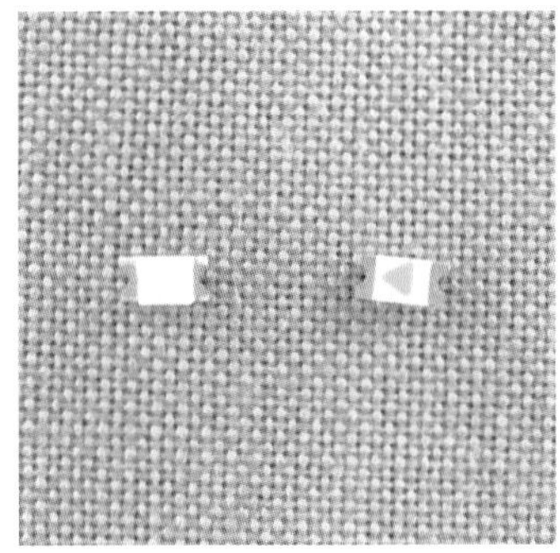

» Tiny, rectangular LEDs with two copper contacts: a positive and a negative. On the back side there is a green arrow or line indicating the negative contact.

» Available in a number of colors, brightness levels, viewing angles, and sizes, SMDs provide the best package in terms of size and brightness.

» The downside is that they are relatively difficult to work with because they are so tiny. But with a little bit of patience and soldering skill, you will be soon incorporating SMDs into your projects.

Ultraviolet (UV) LEDs (aka Blacklight LEDs)

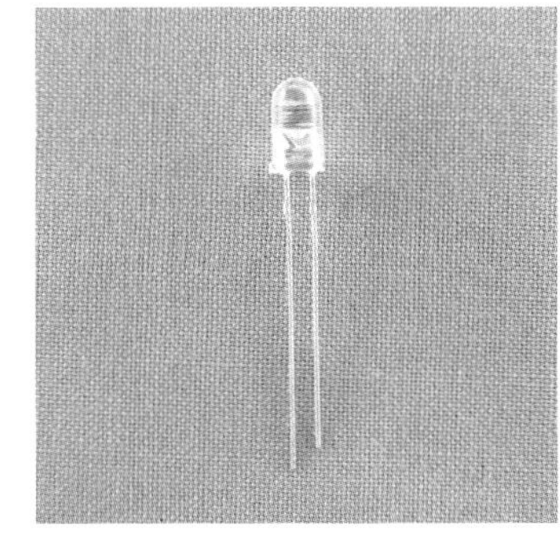

» Emit invisible UV-A rays ranging between 345nm and 400nm.

» When used with fluorescent or phosphorescent materials, produce an interesting light-emitting effect, causing bright objects to glow.

⚠ **UV LEDs are extremely dangerous and cause irreparable damage to your eyes if safety precautions are not taken. You should never handle UV LEDs without proper UV-rated safety glasses.**

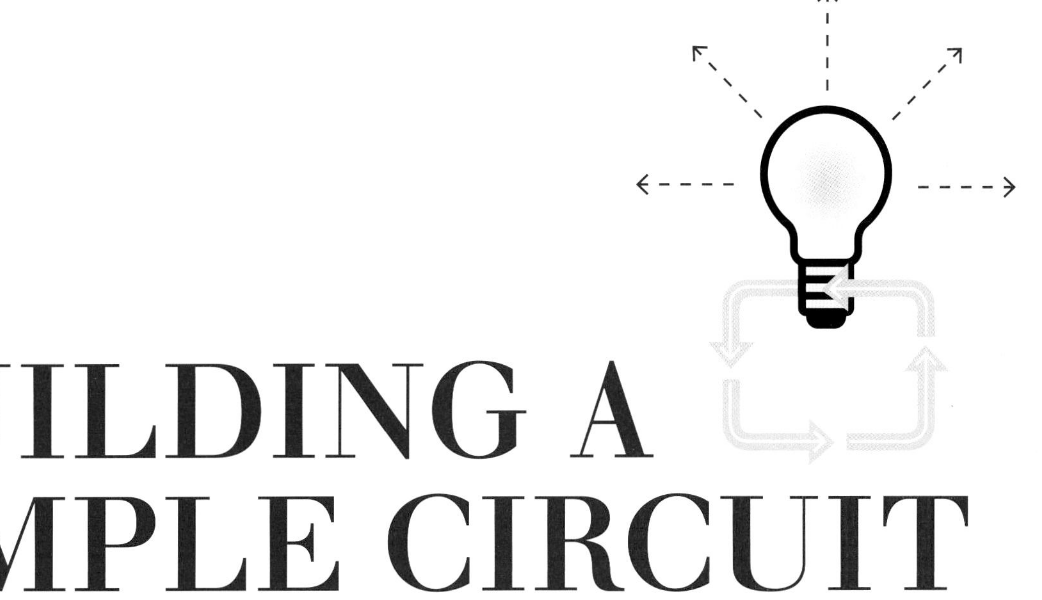

BUILDING A SIMPLE CIRCUIT

By weaving technology — specifically, electronic circuitry — into the objects you make, you can give your projects a nervous system capable of "seeing," "feeling," "reacting," and even "thinking."

Even the most rudimentary electronic circuit has the magical ability to animate the inanimate. Before you begin tinkering with capacitors and diodes, you must first understand a few basic principles behind electronics. With knowledge of the basics, you will be able to create responsive and interactive objects infused with personality and an autonomous life of their own.

The best way to learn and, more importantly, to understand electronics is by creating and experimenting with circuits. The easiest way to prototype or create temporary circuits is by using alligator clips and a breadboard.

The Basics: Voltage, Current, and Resistance

The three most basic concepts in electronics are *voltage*, *current*, and *resistance*. A simple circuit is made from components that use voltage, current, and resistance to perform some function, such as turning on an LED.

VOLTAGE is the driving force in electric circuits. It excites electrons, causing them to move through a circuit. It is measured in volts, symbolized by V.

A few common sources of voltage are batteries, solar cells, generators, and electronic power supplies such as wall adapters. Wall adapters don't actually produce electrical energy, but simply channel and convert it from a wall outlet. These sources of voltage are typically referred to as the "power supply" or "power source" for circuits.

Take a look at a battery, for example. Batteries have a fixed voltage that is usually specified on the battery itself. Every battery has a positive and negative terminal or side, marked respectively by a "+" and a "–". In a circuit, the electrons flow from the negative terminal of the battery, through the circuit, and back to the positive terminal of the battery. Once electrons are energized, they begin to move through a circuit.

CURRENT is the movement or flow of electrons in a circuit. It represents the amount of electrical charge and is what gets work done (for example, turns on a light). It is measured in amperes, symbolized by A. As you will be working with low voltages and currents, the unit of measure you need to become familiar with is the milliampere (mA) which is one-thousandth of an ampere.

RESISTANCE The movement of electrons (current) through a material results in the occasional collision with atoms that restricts their flow. This restriction of electron flow is known as the material's *resistance*. This resistance creates a dispersion of excess energy in the form of heat. Materials with high resistance (plastics, for example) are poor conductors of electricity, and materials with low resistance (metal wires, for example) are excellent conductors. Resistance is measured in ohms, represented by the symbol Ω.

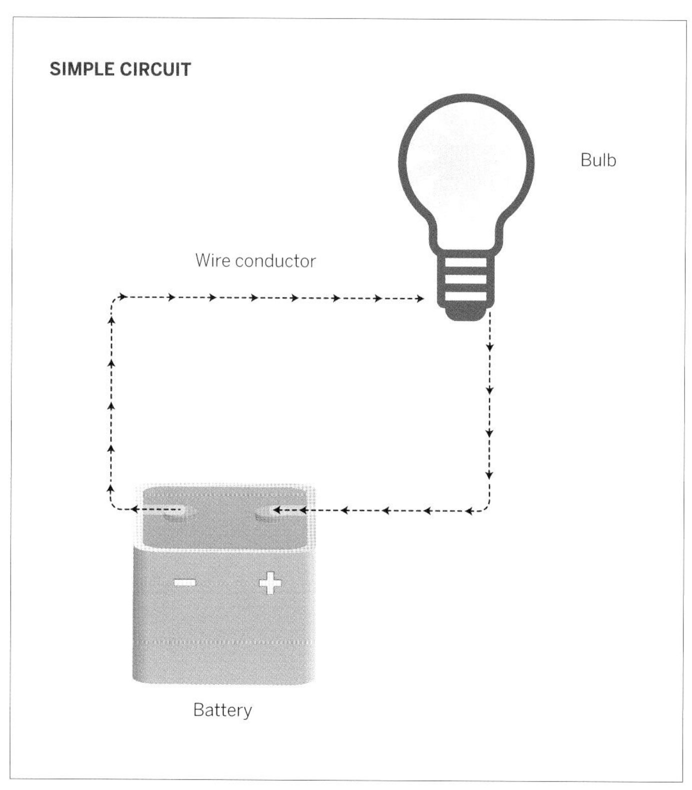

IN SUM You need voltage to jump-start and push the flow of electrons (the current) through a material. Every material has some level of resistance, thereby restricting the current flow.

Prototyping a Simple Circuit Using Alligator Clips

» A basic electric circuit consists of a voltage source providing electrical energy, some device or "load" to make use of that energy, and a path for the current to flow between the source and the load. To understand the basic principles, let's now build a simple circuit using alligator clips.

WHAT YOU'LL NEED

- » **9V battery** the voltage source
- » **Standard red LED** the load
- » **3 alligator clips** that will provide a temporary path for the current to flow between the battery and the LED
- » **390Ω–800Ω, ¼W resistor** to limit the current flow to levels safe for the LED

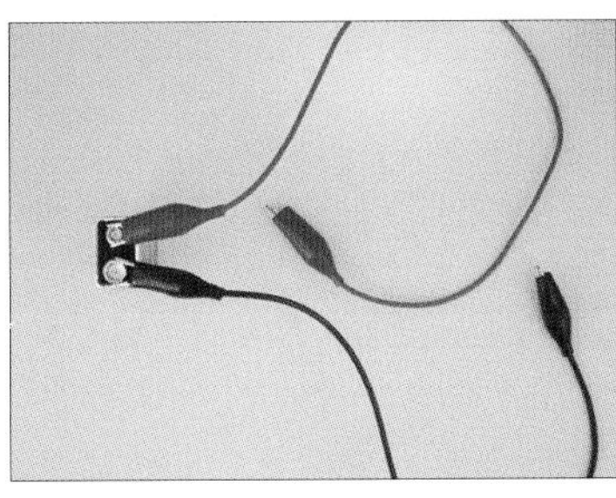

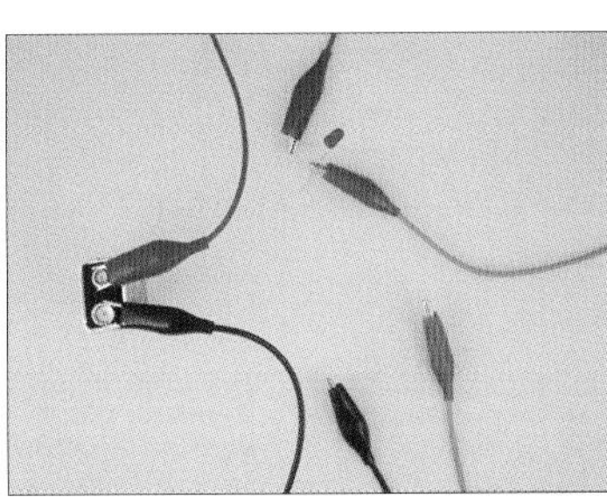

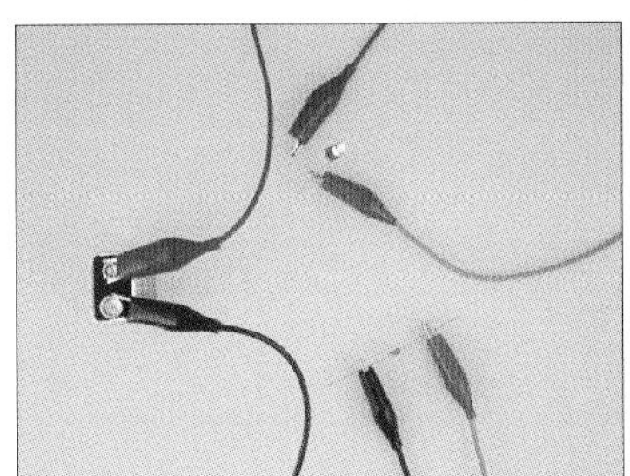

Because you are using a 9V battery, a resistor is absolutely necessary in this circuit to limit the current flow to the LED to avoid overloading and burning out the LED.

STEP 1: Attach a separate alligator clip to the positive and negative terminals of your battery, ensuring that the teeth of the alligator clips do not touch.

NOTE: It is good practice to use different-colored alligator clips to distinguish visually between the positive and negative terminals of the battery. Usually black is used for the negative terminal of a power source and red for the positive.

STEP 2: Grab the alligator clip connected to the positive terminal of the battery. Connect its opposite end to the positive lead of the LED. The positive lead of the LED will be the longer lead.

STEP 3: Using the remaining alligator clip, connect the negative (short) lead of the LED to one side of the resistor.

STEP 4: Grab the alligator clip connected to the negative terminal of the battery. Connect its opposite end to the resistor, completing the circuit. The LED should light up.

NOTE: If the LED doesn't light up, go back and make sure that you have connected the circuit properly. LEDs are polarized, meaning that the current from your power source (the battery) can run through them in one direction only. The current must flow from the positive side of your battery to the positive lead of the LED and back from the negative lead of the LED to the negative terminal of the battery in order for the LED to light up. However, plugging in the LED backwards will not harm it; it merely will not light up.

What exactly is happening in the circuit?
Electrons from the negative terminal of the battery flow through the alligator clips and through the LED, lighting the LED, and back again to the positive terminal. The resistor functions to limit the current to safe levels.

Prototyping a Simple Circuit Using a Breadboard

Once your circuits get more complex, you will need a breadboard to temporarily connect all the electronic components together.

» Breadboards are a quick and easy way to build circuits, but first you need to understand how they work. The easiest way to understand where the holes are connected and where they are not is by removing the plastic housing to see the strips of metal underneath. If you do not own a breadboard with a plastic encasement that can be easily removed, refer to the picture at top right for reference. Although the pattern of holes varies from model to model, the basic arrangement of a breadboard is explained below.

ANATOMY OF A BREADBOARD Underneath the perforated plastic are metal strips (nickel-plated) arranged in rows. The typical breadboard has two columns of many rows of shorter metal strips aligned parallel to one another (A). The two columns are separated by a center divider (B). These two columns are used to connect the electronic components. Along each side is a column of one or two longer strips (C). These outer columns are usually reserved for the power supply.

If you have removed the plastic housing from your breadboard, you can now secure it back in place. The metal strips that you just observed (A) are connected to the holes on top of the board. The holes are spaced 0.1" apart — the standard spacing for the leads of electronic components. The holes that correspond to the metal strip beneath — typically five holes per strip — are connected, and those beside them are not.

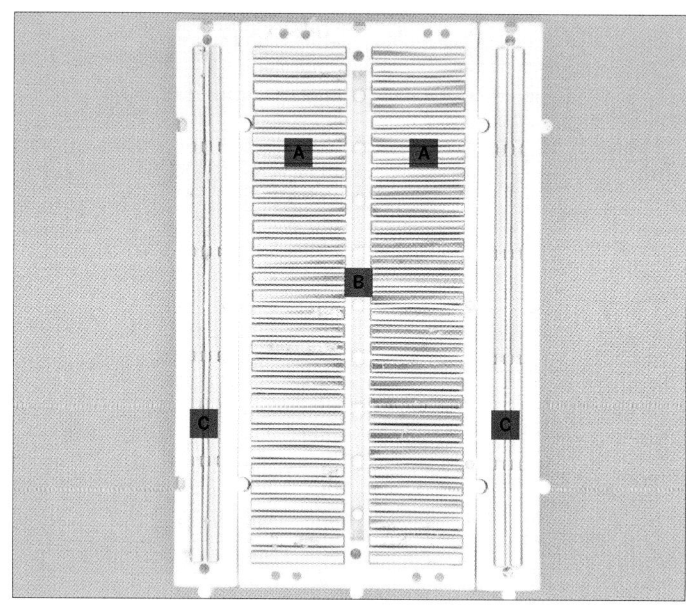

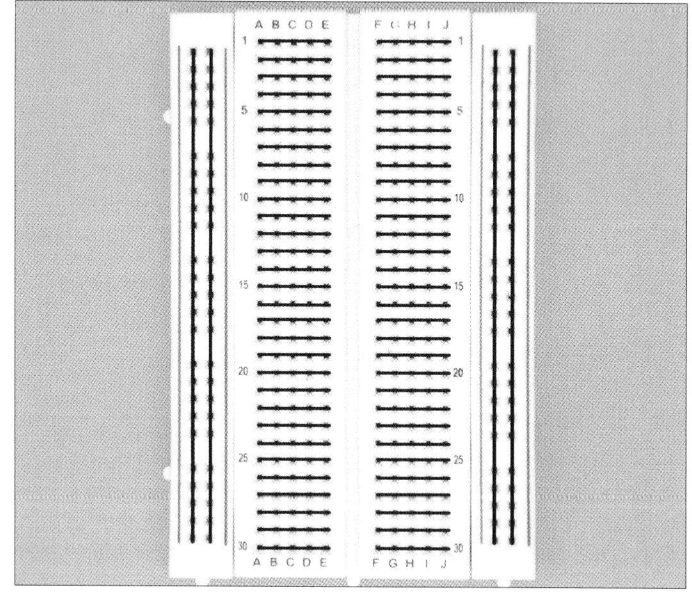

USING YOUR BREADBOARD To use your breadboard, simply push the leads or legs of the components into the holes. To connect two components together, you must either slip their leads into a row of holes connected together or connect them using jumper wires. Jumper wires come with some breadboard kits and are designed in premeasured lengths to fit breadboards. If you don't have jumper wires, you can create your own by cutting and stripping ordinary nonstranded (solid) wire (see how-to at right).

The center divider on a breadboard (B) is used to place integrated circuit (IC) chips, such as the Hex-Schmitt inverter. IC chips are placed in the middle of the board along the divider line so that half the leads are on one side and the other half are on the other. This allows for each lead of the IC chip to be addressed separately and leaves four holes available for components to be connected.

POWERING THE CIRCUIT To power your circuit, you should connect your power supply to the outer columns (C). Your breadboard may have two colored lines, a red and a blue, accompanying each row in the column. These designate the power (red) and ground (blue) for the circuit. If your breadboard doesn't have any colored lines, use the top column as your power and the bottom as your ground. This will help keep you from confusing the two and shorting your circuit. To learn more about shorts, see the sidebar at right.

How to Strip Wire

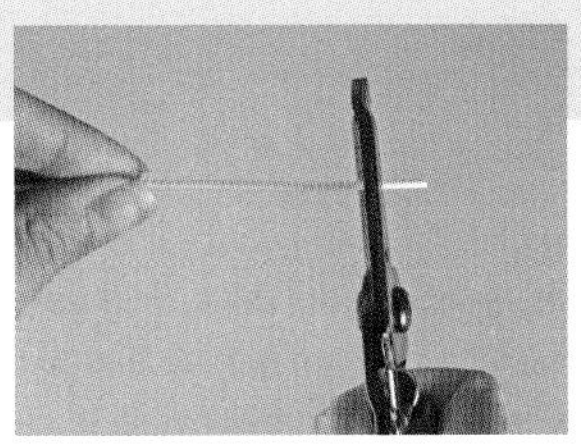

1. Using wire cutters, cut the wire to the desired length.

2. Using wire strippers, remove ¼" of plastic sheath from both ends of the wire.

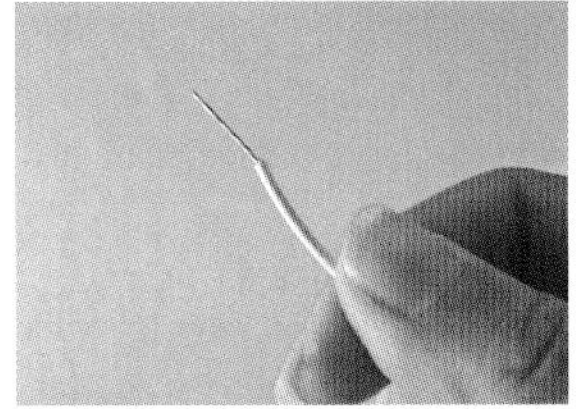

3. For stranded wired, twist the stripped ends together and *tin* the ends to prevent the wire from fraying.

   To tin the ends, add a touch of solder to the stripped wire, coating the ends with a thin layer of solder.

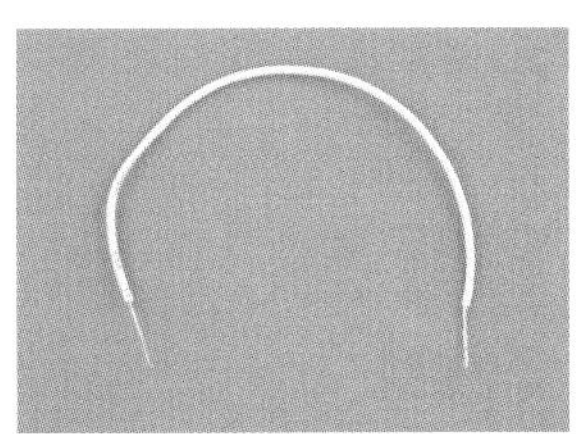

What is a short?

A short occurs when a circuit is wired improperly and the current flows through the circuit along an unintended path. When a short happens, the circuit may not work properly or may not work at all. Shorting a circuit can result in damaged electronic components if an accidental surge of high current flows through them. It is particularly dangerous to short a battery, as the battery will overheat and may explode. If any component heats up unexpectedly, quicky disconnect the power source.

Build a Simple Circuit with a Breadboard

» Now let's tinker with the breadboard. We are going to build a similar circuit as we did before, but this time with the addition of an extra LED.

WHAT YOU'LL NEED

- » **9V battery and battery connector**
- » **Needlenose pliers**
- » **Jumper wires**
- » **2 LEDs** of the same color
- » **Breadboard**
- » **390Ω–800Ω, ¼W resistor**

STEP 1: Place the longer, positive lead of the LED in a row above the center divider. Place the shorter lead in another row below the divider. The LED should straddle the center divider of the board.

STEP 2: Using needlenose pliers, gently bend the leads of your resistor into right angles. This will make it easier to place the leads of the resistor into the breadboard.

STEP 3: Connect one lead of the resistor to the negative lead of the LED, placing it in the same row. Slip the second lead into a different row. Using a jumper wire, connect the second lead of the resistor to the bottom negative column of the power supply.

STEP 4: Using a jumper wire, connect the positive lead of the LED to the top, positive column of the power supply.

STEP 5: Place the battery connector onto the 9V battery. Slip the positive and negative leads into the top and bottom columns, respectively. The LED should light up.

STEP 6: Now try adding a second LED to the breadboard. You can connect the LED in either parallel or series (page 22).

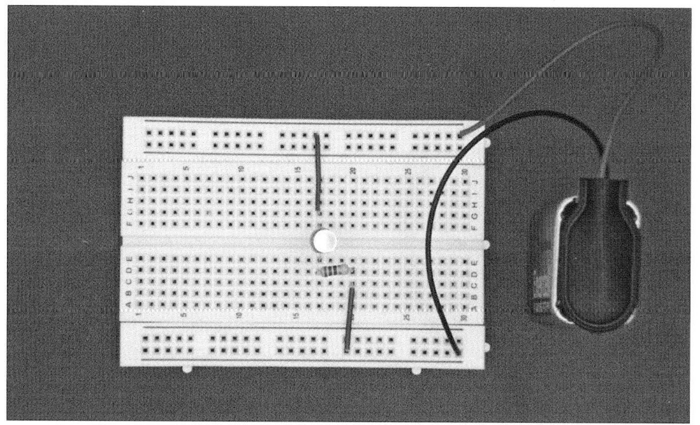

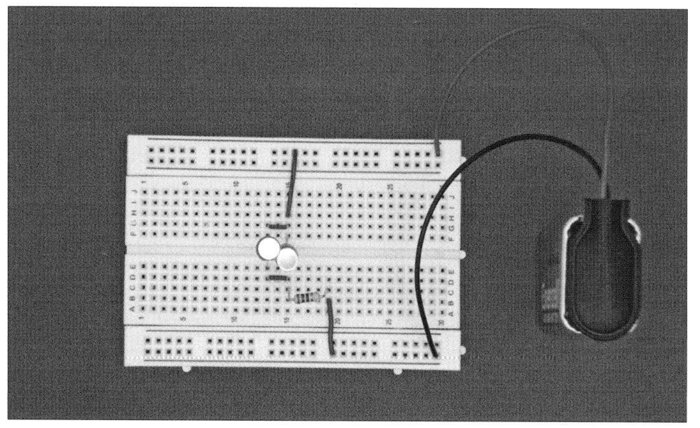

Prototyping a Circuit Using a Perforated Board

Perforated boards, typically referred to as perfboards, are pre-punched boards with or without copper traces, used to prototype circuits. Unlike with breadboards, electronic components are soldered to each other or onto the board itself, making the circuits permanent.

» There are two types of perfboards: boards with copper traces and those without. Perfboards without copper traces are generally less expensive but they are more difficult, and messier, to work with. On these boards, the leads of electronic components must be soldered to each other directly, or by using wires to connect components to one another.

For a much cleaner and simpler alternative, use perfboards with copper traces, specifically ones laid out with IC spacing. Perfboards with IC spacing allow you to easily transfer a circuit built on a breadboard directly to a more permanent circuit soldered onto a perfboard.

» Perfboards with IC spacing work similarly to breadboards. The leads of electronic components must be bent and inserted into the through-holes from the top of the board (the side without the copper traces). The leads then can be bent flush to the bottom of the board to hold the components temporarily in place. Similar to using a breadboard, jumper wires come in handy to jump components to different rows or to connect different components to each other. Once you have all the components in place, you can solder the leads of each component to the copper pads underneath the board accordingly. Once all the components are soldered, the excess leads can be trimmed using a pair of wire cutters.

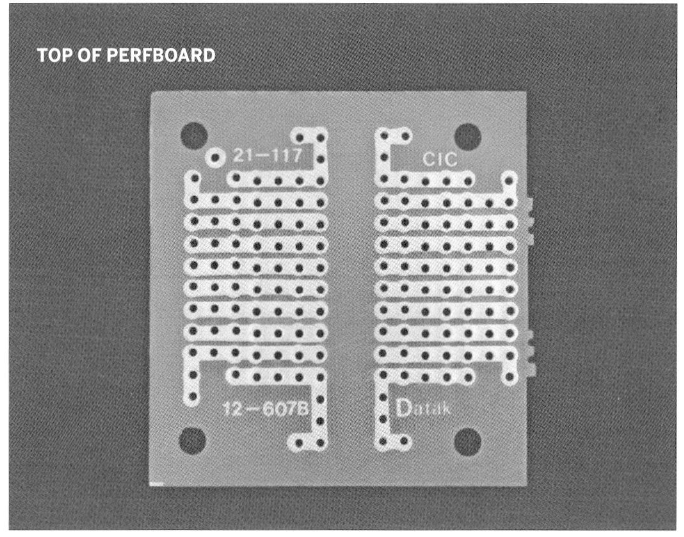

TOP OF PERFBOARD

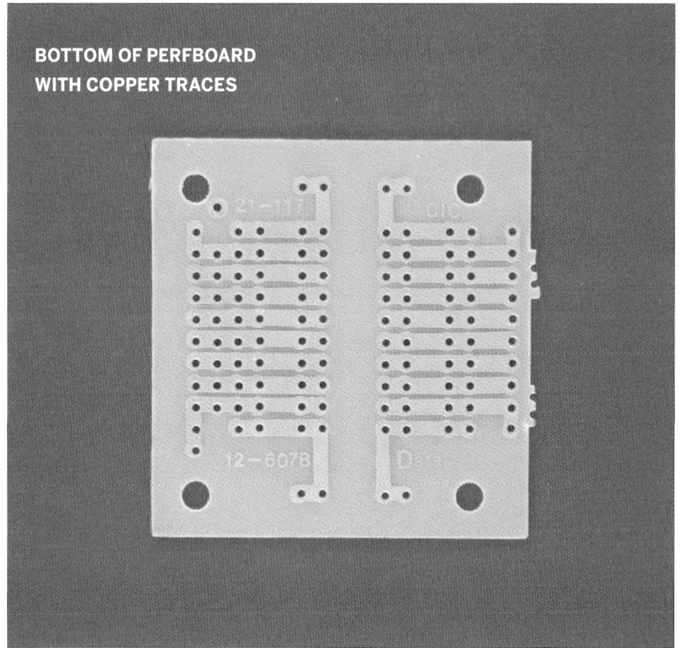
BOTTOM OF PERFBOARD WITH COPPER TRACES

CHOOSING THE BEST POWER SUPPLY

Your power supply is the life source of your project. When you begin designing your project, your choice of power source will affect not only the design of your circuit but the form of your project as well. Choosing a power supply is determined by technical and economic factors as well as design decisions. Is your project small and portable? How many hours does it need to work before it can be recharged or the power source replaced? Is the project used predominately outdoors, indoors, or both? Taking the variables of size, duration, portability, and environment into consideration will help you determine the best power sources for all your projects.

Because you will be working exclusively with low voltages in the projects in this book, the three main sources of power that you should become familiar with are batteries, solar cells, and low-voltage power adapters. Following is an explanation of the three, including the benefits and disadvantages of each.

Understanding Batteries

» For the majority of your projects, batteries will be your primary power source. They are portable; available in different sizes, shapes, and weights; and some are rechargeable. The major downside of batteries is that they are not very sustainable. If you don't choose your battery taking into consideration the amount of power required and consumed by your project, you may find yourself having to recharge or replace your batteries often.

The two most important units that you need to understand in batteries are their voltage and capacity ratings.

VOLTAGE The predominate factor in choosing a battery is its voltage. In your projects, you must use the minimum amount of voltage required by your circuits to get them to work properly (see page 28 for more).

CAPACITY Although batteries have a fixed voltage, their capacity is variable. If you think of your battery as the storage container for electrical energy, the amount of electrical energy available for use in a battery over a period of time is its capacity, expressed in ampere-hours (Ah). In general, the higher the ampere-hour rating, the longer the battery will last for a certain load. For example, if you have two 3V lithium batteries rated at 250mAh and 500mAh, the second battery should theoretically last twice as long when used in the same circuit.

So why not always use a battery with more capacity? This is where economics, battery size, and weight come into play. Generally, when comparing batteries with the same composition and voltage but different capacities, the battery with a greater capacity will be heavier, larger in size — and probably more expensive.

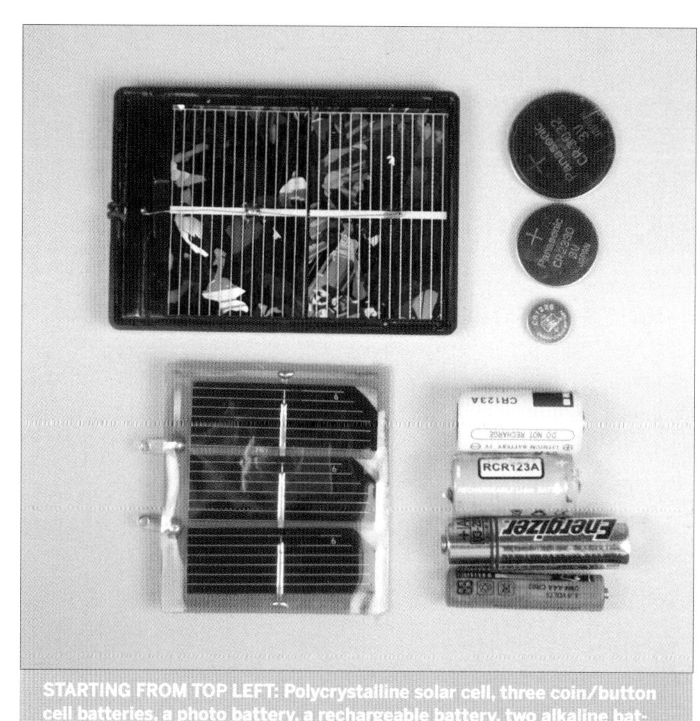

STARTING FROM TOP LEFT: Polycrystalline solar cell, three coin/button cell batteries, a photo battery, a rechargeable battery, two alkaline batteries, and a monocrystalline solar cell.

Connecting Batteries: Series Versus Parallel

Most electronic gadgets and toys use more than one battery. Similarly, in your circuits, you may choose to group batteries to get the desired voltage and capacity.

» Batteries can be grouped together in series for higher voltages, or in parallel for more capacity.

TO GROUP BATTERIES IN SERIES The positive terminal of the battery is connected to the negative terminal of the next battery, and so on, as illustrated at far right. The result is an increase in voltage that is the sum of the individual battery voltages, while the capacity remains the same.

FOR EXAMPLE If you group four 3V batteries with a 200mAh capacity, the result is a 12V power supply with a capacity of 200mAh.

TO GROUP BATTERIES IN PARALLEL All positive terminals are connected together and all negative terminals are connected, as illustrated at far left. In a parallel connection, the capacity is increased to the sum of the individual battery capacities, while the voltage remains the same.

FOR EXAMPLE If you group four 3V batteries with a 200mAh capacity, the result is a 3V power supply with a capacity of 800mAh.

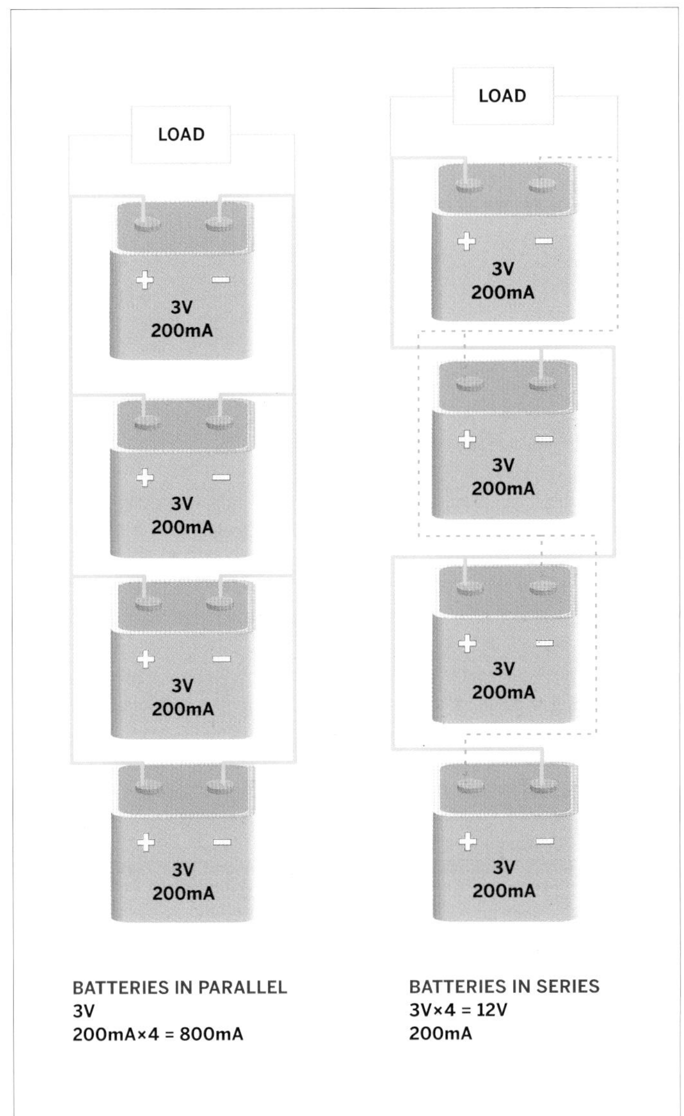

BATTERIES IN PARALLEL
3V
200mA×4 = 800mA

BATTERIES IN SERIES
3V×4 = 12V
200mA

Connecting Solar Cells: Series Versus Parallel

Similar to batteries, solar cells can be grouped into solar panels connected in series or parallel.

» Solar cells grouped together in series provide higher voltages, and when grouped in parallel, provide more capacity. Unless you purchase solar panels, most individual solar cells or economy packs of broken solar cells don't come with leads presoldered to their positive and negative sides. You must either solder them (refer to page 41) or glue them on yourself with conductive epoxy. You will learn more about solar cells on the following pages.

Different Types of Batteries

There are plenty of batteries to choose from, ranging in composition, size, shape, voltage, capacity, and weight.

There are two different classes that batteries are divided into: *primary* and *secondary*.

PRIMARY BATTERIES are disposable. They are used once and discarded when they are drained.

SECONDARY BATTERIES can be recharged and reused many times.

Without getting into the nuances of various battery compositions, the following is a general overview of the different battery types and their characteristics to help you determine the most appropriate power supply for your projects.

Alkaline Batteries

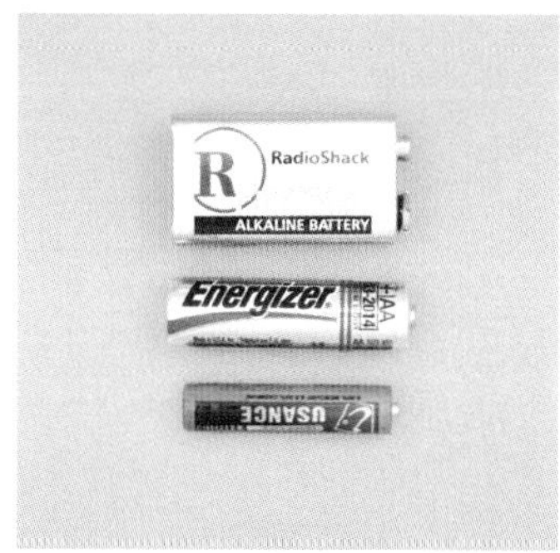

» The most common primary batteries used in electronic devices.

» They come in five different sizes: AA, AAA, C, D, and 9V. All except the 9V are cylindrical in shape and rated at 1.5V. The 9V is square, and as its name suggests, is rated at 9V.

» Their capacity increases with battery size, AAA being the smallest and D the largest.

Advantage: Ubiquitous and therefore inexpensive.

Disadvantage: Size, weight, and low voltage. Not rechargeable.

Lithium Polymer/ Cellphone Batteries

» Secondary flat, rectangular batteries used in a variety of portable devices such as cellphones, PDAs, and cameras.

» Rechargeable, fairly lightweight, and compact, with a high capacity rating.

THINK TWICE about throwing away or recycling your old cellphone — it may be the best and cheapest power supply for your projects.

Advantage: Ideal for any mid-sized portable project that requires a great consumption of power.

Disadvantage: Typically only available at 3.7V — the standard voltage for cellphones. Commercial battery contacts, battery holders, and battery chargers are not generally available for these batteries, so a custom one must be designed. Lastly, they are quite expensive.

Photo Batteries

» Similar to alkaline batteries in their cylindrical shape and appearance, but rated at 3V, 6V, and even 12V.

» Typically used in cameras.

Advantage: Voltage. You only need one photo battery versus two alkaline to get the same amount of voltage.

Disadvantage: Cost. One photo battery typically still costs more than two alkaline batteries.

The Alternative Energy Source

Rechargeable Batteries

» The most sustainable and environmentally friendly option.

» When drained, can be restored to full charge and reused.

» Available in most common battery sizes, including coin cell.

Advantage: Over the long term, they end up saving you money and saving the environment.

Disadvantage: Initial investment is costly. The batteries themselves are expensive and require an accompanying battery charger. Also, secondary batteries generally have a lower capacity than primary batteries, and some even require special disposal.

Watch or Coin/Button Cell Batteries

» Primary small, circular batteries commonly used in watches, calculators, toys, and so on.

» Typically available from 1.5V–3V.

» An excellent choice in projects with a light load that require a compact, lightweight power source.

» Their size is delineated in millimeters (mm), specifying the outer diameter of the battery.

» Secondary coin cell batteries are also available, but are not very economical.

Advantage: Compact size.

Disadvantage: Fairly limited in capacity.

Solar Cells

» Solar cells, or *photovoltaic cells*, make use of the greatest renewable energy source — the sun — by converting light energy into electrical energy.

» The most sustainable power source available; well suited for wearable and outdoor projects.

Advantage: Solar cells make excellent light sensors. They distinguish between areas of light and dark and different times of day. Solar cells allow your projects to be aware of the fluctuating lighting conditions of the environment, and with the help of a few electronic components, this awareness can be easily translated into an infinite number of behaviors. They are flat and lightweight. Some are even flexible, and can easily be bent and curved to fit 3D surfaces.

Disadvantage: They don't work at night without reserved energy. They work poorly in environments with low outdoor lighting conditions. They are not suitable for applications that require a consistent and continual power source. However, with the addition of a few electronic components, electrical energy generated by solar cells can be stored and used at night. Finally, solar cells are still fairly expensive, and you need quite a few of them to generate a notable amount of power. For projects with a light load, this isn't too much of a concern.

Different Types of Solar Cells

Here is a general overview of the different solar cell types and their characteristics to help you determine the most appropriate solar cell for your projects.

There are few different types of solar cells available on the market. All solar cells share similar designs, and offer various advantages and disadvantages, depending on the materials used in their construction.

Amorphous

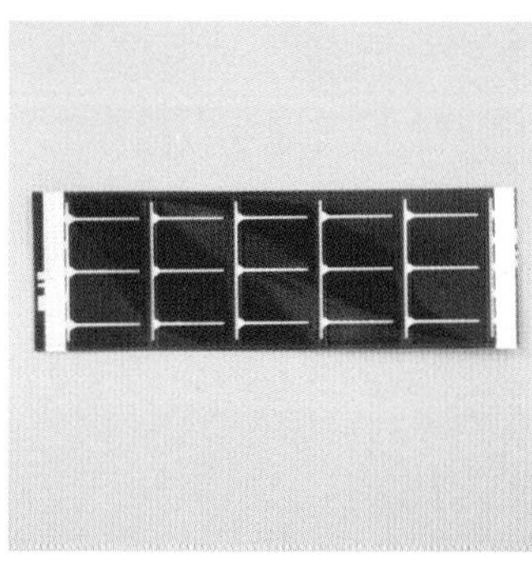

» Made from placing a thin film of active silicon on a solid or flexible backing.

» These are the least efficient of all three types, but they offer a level of flexibility that the other two do not.

Monocrystalline

» Produced from a thin slice of a single silicon crystal.

» The most efficient, and typically the most expensive.

Polycrystalline

» Produced from a thin slice of a cast silicon block, they have the appearance of blue shattered glass.

» Cheaper and less efficient than monocrystalline solar cells.

Soldering Leads onto Broken Solar Cells

» Soldering leads onto solar cells can be a bit tricky. First, you need to determine the positive and negative sides of your solar cell. The back (typically gray) side of the solar cell is the positive side; the front (blue or black, depending on what type of solar cell you have) with the lines is the negative.

WHAT YOU'LL NEED

- Solar cells
- Fine-grain sandpaper
- Soldering iron and solder
- Thin gauge, nonstranded wire

STEP 1: Using a piece of fine-grain sandpaper, carefully sand a small section of the thicker metallic bar running perpendicular to the thinner lines on the front, negative side of your solar cell. The section should be smooth and shiny.

STEP 2: Set your soldering iron to low temperature. With the tip of your soldering iron, heat the section you have just sanded for a second or two. Then add a touch of solder, drawing it along the bar.

NOTE: If your solder balls up, either your iron is too hot or the metallic bar hasn't been sanded properly.

STEP 3: Cut your lead wire to the desired length and strip ½" off both ends. Be sure to use a thin-gauge, nonstranded wire. Place one of the stripped ends of your wire directly onto the solder contact that you created. Use the tip of your soldering iron to reheat the solder and melt the solder over the wire.

STEP 4: After you have soldered on your negative lead, flip the cell over to solder on the positive lead. The back side of the solar cell does not need to be sanded.

STEP 5: Cut and strip another piece of wire. Create a solder contact by heating the cell for a second and adding a drop of solder. Attach your wire to the solder contact.

STEP 6: Repeat for the other cells you want to connect.

NOTE: If your wires won't adhere to your solar cells, you can use conductive metal tape to create contact points to solder onto. Another option is to use conductive epoxy (see page 4) to glue the wires onto the solar cells.

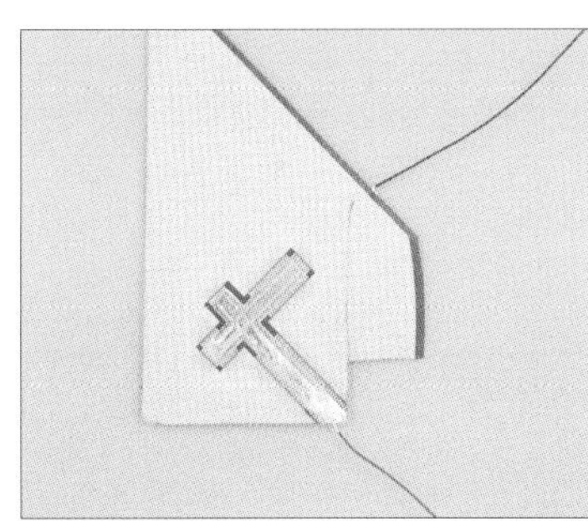

Power Adapters

Power adapters convert AC (alternating current) voltage from a wall outlet to a constant DC (direct current) voltage (what you need to power your projects). They come in different voltage outputs, current outputs, power ratings, and various connector types and sizes.

BASIC EXAMPLE Take a look at your cellphone power adapter. The ratings vary from model to model, but typically the output will be within the range of 3.7V–5V at 400–700mA of current. Your cellphone power adapter converts the 120V AC (common in the U.S.) coming from your wall outlet to safe operating levels for your phone.

» The main limitation to all batteries is that they eventually drain. If you need a continuous power supply for your projects, your only option is to use a power adapter.

FOR EXAMPLE If you are creating an LED desk lamp or indoor night light, you need a power source that will last for hours. Batteries, in this case, are not an economical or logical option.

» Besides choosing an adapter with the appropriate voltage and current output, you have a choice of purchasing regulated or unregulated power adapters. Unregulated power adapters are significantly less expensive than regulated adapters, because they do not have a voltage regulator built into them. What this means is that they are designed to provide a specific voltage at a specified load.

FOR EXAMPLE An unregulated 3V power adapter with a 500mA current rating will provide 3V if the circuit draws close to 500mA of current. If your circuit uses less current, the voltage may jump up and vice versa. An increase in voltage to a circuit can damage the components in that circuit.

» Unless you are certain that you know the exact current draw of the circuit, use a regulated power adapter. A regulated power adapter supplies the exact voltage specified, regardless of the amount of current consumed, as long the rated output current is not exceeded.

TUTORIAL

Hacking into a Power Adapter

» Most power adapters on the market come with some type of connector. To incorporate the power adaptor in your project, you must first remove the connector. Second, you need to distinguish the polarity of the two wires.

WHAT YOU'LL NEED

- Multimeter
- Electrical tape
- Wire cutters
- Wire strippers
- Power adapter

Unfortunately, there is no standard or conventional marking on the plastic coating of the wires to help determine the positive from the negative. Even with two similar wall adapters using the same visual conventions, there is no guarantee that the wire is the same in both. The best way to distinguish the polarity of the wires is by using a multimeter (see page 70). To determine the polarity of your power adapter, follow the next four steps.

STEP 1: Clip the connector off the wall adapter and separate the two wires. Cut one of the wires slightly shorter than the other.

STEP 2: Using wire strippers, strip about ¼" of insulation from both wires.

STEP 3: Using electrical tape, tape the two wires apart from each other onto a hard surface such as a table.

⚠ **STEP 4:** Now plug the wall adapter in.

STEP 5: Set your multimeter to the voltage setting and select the appropriate range for the voltage being measured. Using the red (positive) and black (negative) probes of your multimeter, hook the probes randomly to the wires to see if the voltage measured is negative or positive. If you've never used a multimeter before, see page 70. If it is negative, reverse the leads. Now you know that the wire that the black lead is attached to is the negative (ground) wire, and the other is the positive wire.

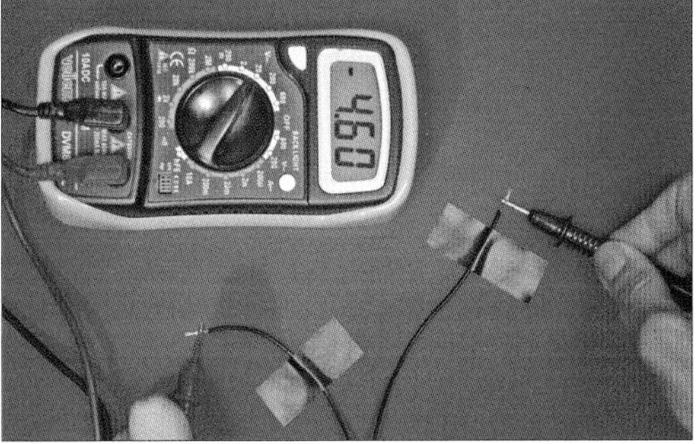

« If the voltage reads negative, switch the probes to get a positive voltage.

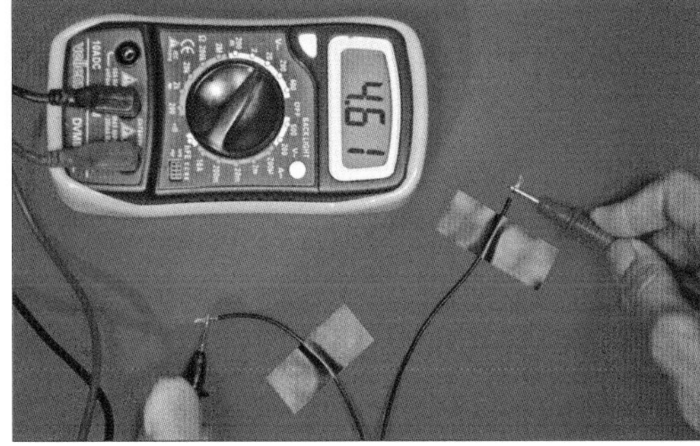

« If the voltage reads positive, the wire attached to the black probe is the negative wire, and the wire attached to the red probe is the positive.

⚠ **ELECTRICITY SAFETY** Once plugged in, do not touch the stripped wires with your fingers to prevent shocking yourself. The wires also should not touch each other to prevent from shorting and ruining your wall adapter.

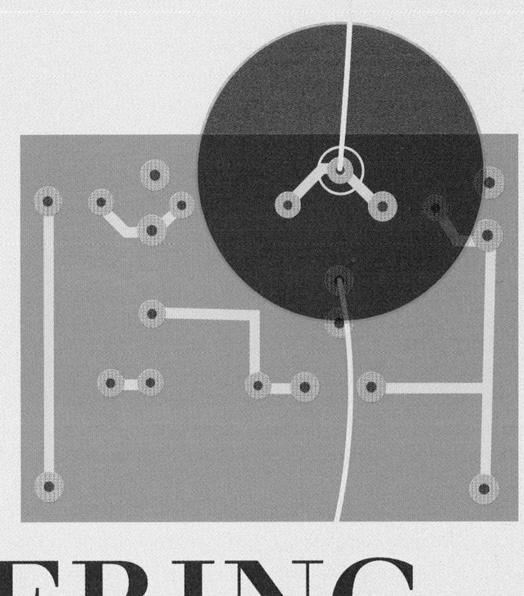

THE ART OF SOLDERING

The two most fundamental skills needed to assemble the electronic projects covered in this book are soldering and sewing. Soldering is the traditional way of joining electronic components to form an electrical connection. It is a delicate skill that is not difficult to master once you understand the tools you are using and the basic concepts. The art of making a perfect solder joint takes some practice but, just like riding a bike, easily becomes second nature.

Weak or faulty solder joints may result in needless hours spent troubleshooting your circuits or, ultimately, your circuits malfunctioning entirely. With a little a patience, practice, and the right tools, you will be making shiny joints in no time.

The Tools

SOLDERING IRON/STATION There are many soldering irons on the market to choose from that range vastly in price depending on the features they offer. Soldering irons come in different voltage and wattage ratings, and some have temperature control capabilities.

» The two main features to take into consideration when purchasing a soldering iron are the *wattage* and the *temperature control*. You'll want to purchase an iron rated between 25W–45W. This low- to mid-range wattage works well for soldering electronic components. Using a high-wattage soldering iron (70W for example) increases the chances of damaging an electronic component by adding too much heat. Low-wattage irons (15W for example), on the other hand, increase your chances of making a weak solder joint.

» The more expensive soldering irons come with a temperature control feature that allows you to vary the temperature at the tip. This is ideal if you are going to be working with components that are particularly sensitive to heat.

» Generally, your soldering iron will come with an iron tip. If your iron comes with a copper tip, replace it with an iron tip. Copper tips are more difficult to work with because they corrode quickly.

ROSIN-CORE OR LEAD-FREE SOLDER The solder recommended for use in your projects is 60/40 rosin-core solder. Basic electronics solder is composed of an alloy made of tin and lead. The most common composition is 60/40 solder — 60% tin and 40% lead — which is suitable for all general-purpose electronic projects.

» Typically, solder is in wire form and is sold in spools. Wire solder comes in various diameters. Use a thin solder (0.015"–0.025" diameter) to give you more control over the amount applied. Wire solder contains rosin flux in the center; therefore, unlike in jewelry making, there is no need to use additional flux.

» If you are averse to products containing lead, lead-free solder is available. Lead-free solder contains only tin and silver, but it is very difficult to work with because it has a much higher melting point.

» Always pull or tear solder apart with your hands instead of cutting it. Cutting the solder can cause the rosin to leak out, while pulling it seals the rosin in.

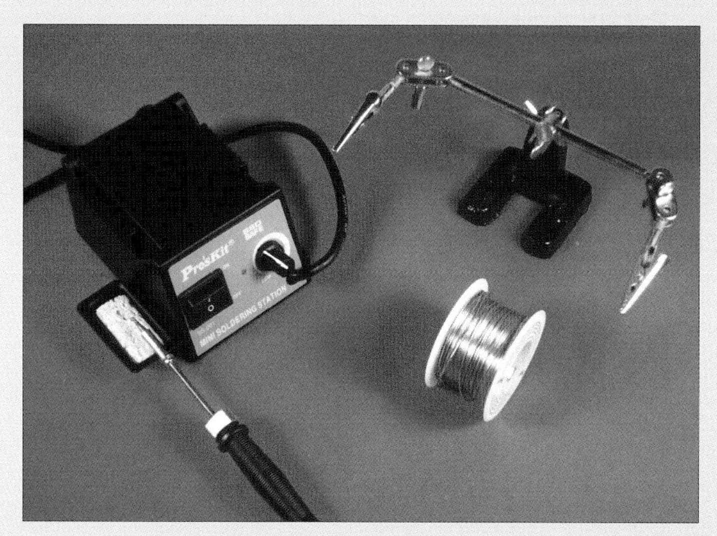

DAMP SPONGE Most soldering stations come equipped with a small sponge. This sponge is dampened with a bit of water and used to clean the tip of the soldering iron. If your soldering station doesn't come with a sponge, you can cut and use a small portion of an ordinary kitchen sponge.

THIRD HAND AND MODELING CLAY The final tool recommended, although not required, is the *third hand* (see page 14). The third hand is an inexpensive, indispensable tool that makes soldering components much easier. This tool is ideal for soldering components to circuit boards, but a bit more clumsy when you need to solder small, delicate electronic components together. In these cases, modeling clay is a much better alternative. Modeling clay can be used to hold electronic components in place temporarily while you solder the components together.

Soldering Basics

» LESS IS MORE

As a rule of thumb, when it comes to soldering, less is more. You want to use just enough heat to heat the joint and melt the solder and the minimum amount of solder to melt evenly over the entire joint. It's easy to tell when you've created a strong joint. A good solder joint will be smooth and shiny, not dull and pitted.

» CLEANING AND TINNING YOUR TIP

Another important factor is cleanliness. A dirty soldering tip or rusty electronic components are difficult to solder, because the solder simply won't adhere to the surface; instead, it will bead into tiny globules. Before you begin to solder, make sure that the tip of your soldering iron is clean and "tinned." To tin the soldering iron, add a touch of solder to the tip and then wipe the excess on the damp sponge. Your tip should be smooth and shiny.

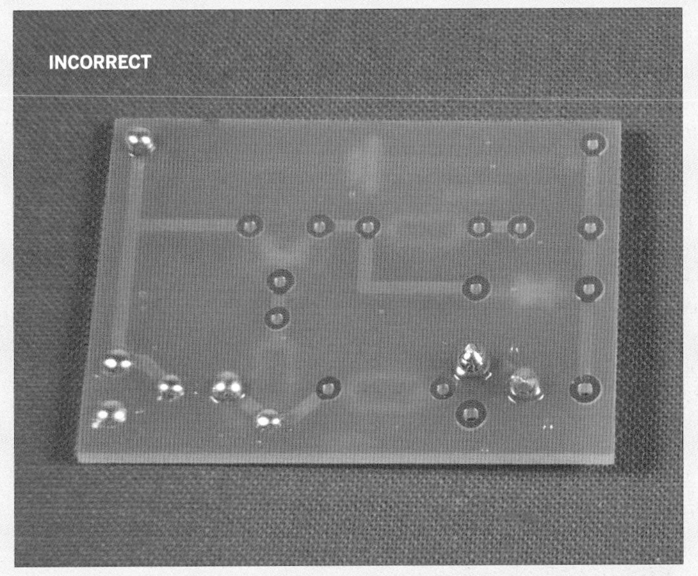

INCORRECT

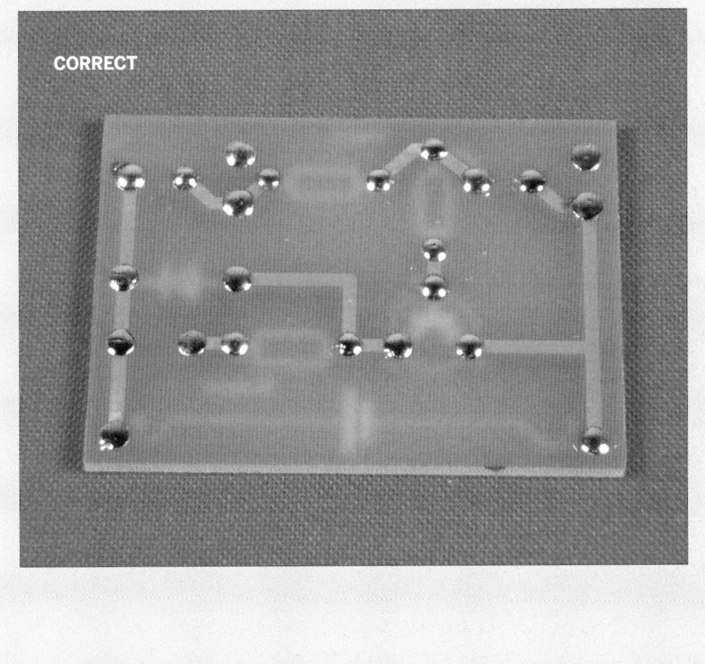

CORRECT

Soldering

» Once you've tinned the tip of your soldering iron, you are ready to begin soldering. Soldering is accomplished in three primary steps: heating the joint area, applying the solder, and removing the iron.

WHAT YOU'LL NEED

- Soldering iron
- Solder
- Protoboard or circuit board
- Electronic components

STEP 1: Heat the joint with the tip of the iron for a few seconds before you apply the solder. Heat must be applied simultaneously to both surfaces being joined in order for the solder to make a proper connection.

STEP 2: Next, apply a touch of solder to the components, not the iron. Generally, you can add heat to one side of the joint while applying solder to the other. The solder should melt, completely covering the entire joint. Remember, less is more. You do not need to add very much solder to create a strong connection.

STEP 3: Remove the iron, allowing the joint to cool. Make sure that the components you have soldered do not move when the joint is cooling. This can result in a weak internal connection that is not visible from looking at the joint.

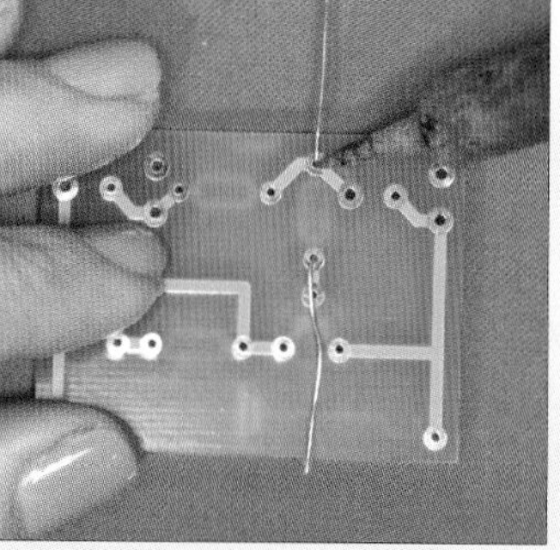

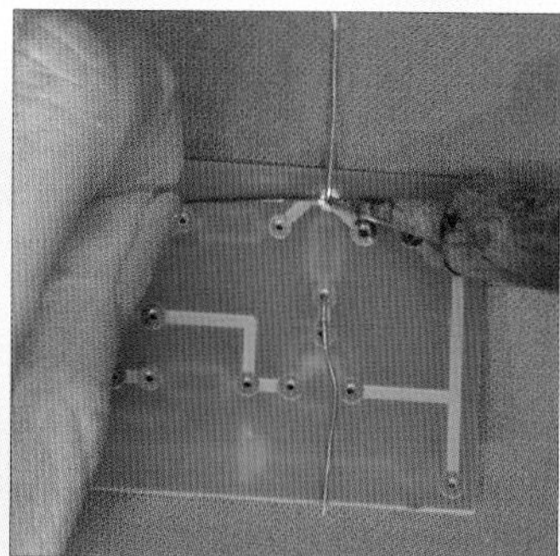

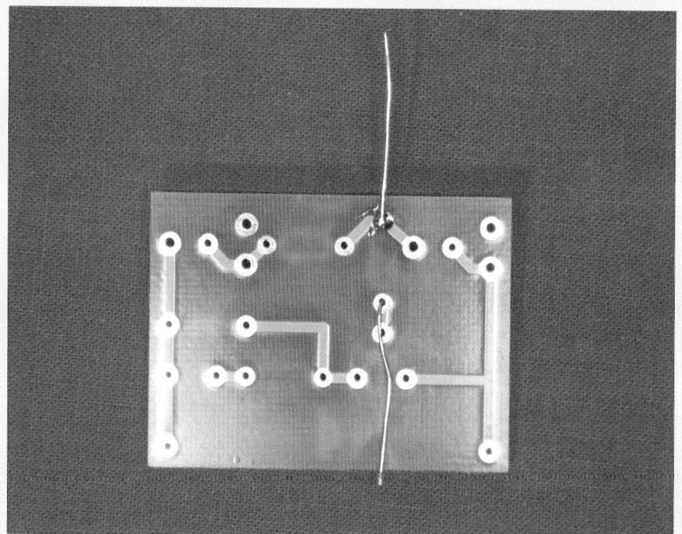

PRACTICE

A good way to practice and fine-tune your soldering skills is by using a copper-etched perfboard and some cheap components. With some practice, you will quickly learn how long to heat your joint area and how much solder to apply to create a perfect, shiny solder joint.

Desoldering

WHAT YOU'LL NEED

» Soldering iron
» Desoldering braid

» Whether you are scavenging parts from other electronics or needing to resolder a messy joint, there will undoubtedly come a time when you need to remove solder from a joint.

To desolder a joint you will need to have a desoldering braid handy. A desoldering braid is a fine copper braid that draws molten solder into its mesh, removing it from the components. Follow these four steps to desolder a joint:

STEP 1: Place the desoldering braid over the solder joint that needs to be removed. **CAUTION: Only hold the braid using the tube it comes in. The braid will get extremely hot.**

STEP 2: Using the tip of the iron, press the tip over the braid and joint. Once the solder melts, it will be absorbed by the braid. Be careful not to overheat the joint and the components.

STEP 3: Remove the braid while the solder is still molten; otherwise, the braid will stick to the components.

STEP 4: Repeat these steps with a clean piece of desolder braid if necessary.

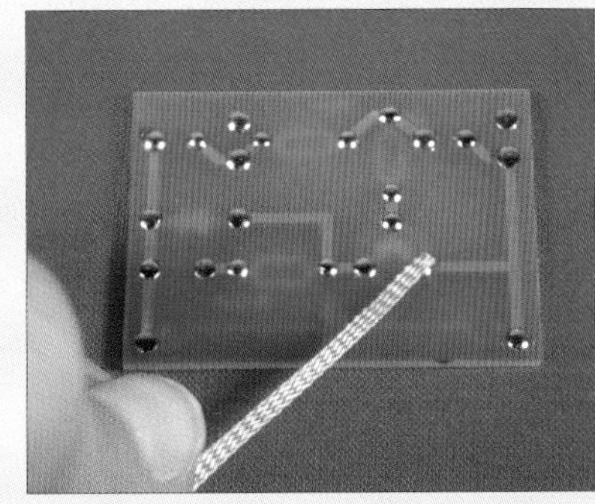

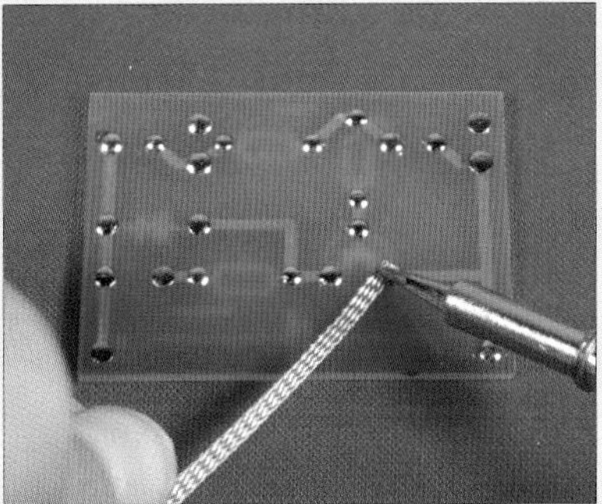

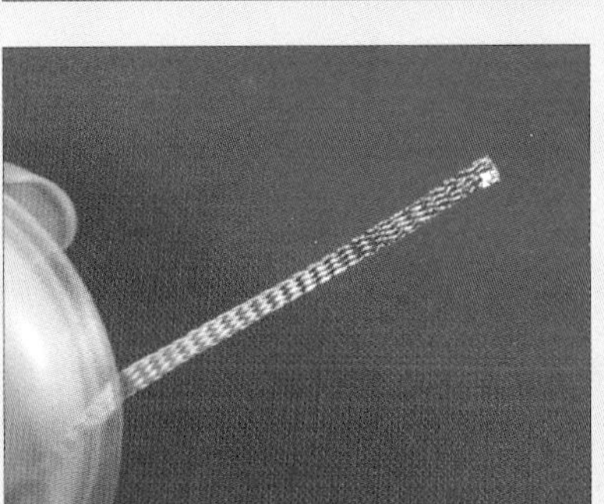

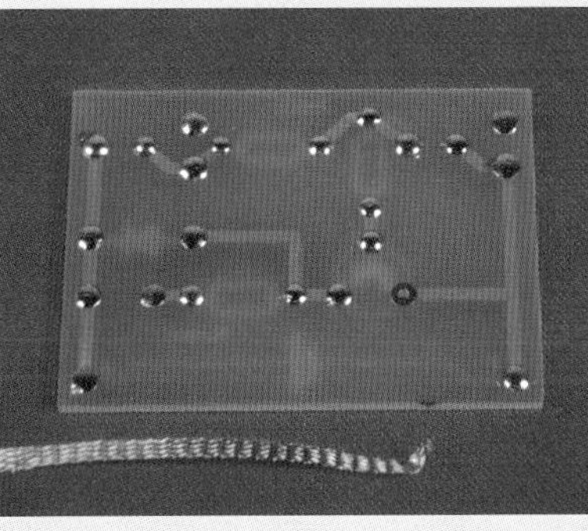

SOLDERING SAFETY

1. **Never touch the soldering iron. The soldering iron is going to become extremely hot and will burn you. Always return the soldering iron to its stand.**

2. Wear safety goggles or eye protection when soldering. **Loose solder (hot molten metal) on the tip can easily be projected into your eyes and face. DON'T EVER FLICK EXCESS SOLDER OFF THE TIP. Use the sponge to clean and remove excess solder.**

3. Work in a well-ventilated area. **The fumes from the solder are toxic and can be harmful and irritating.**

4. Wash your hands after you solder and never eat while soldering. **Solder contains lead, and lead is poisonous.**

THE BASICS OF SCREEN PRINTING

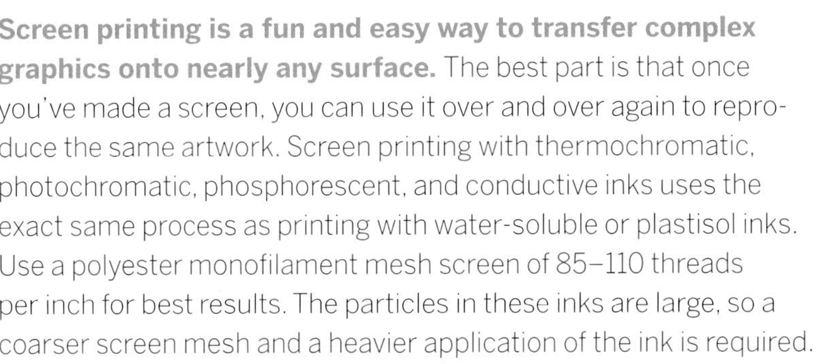

Screen printing is a fun and easy way to transfer complex graphics onto nearly any surface. The best part is that once you've made a screen, you can use it over and over again to reproduce the same artwork. Screen printing with thermochromatic, photochromatic, phosphorescent, and conductive inks uses the exact same process as printing with water-soluble or plastisol inks. Use a polyester monofilament mesh screen of 85–110 threads per inch for best results. The particles in these inks are large, so a coarser screen mesh and a heavier application of the ink is required.

Phosphorescent inks typically require you to mix a phosphorescent pigment with an acrylic or water-soluble base before you begin printing. Dry cleaning or machine drying any of these screen-printed fabrics is not recommended, as they will lose their transformative properties.

Choice in Materials

CHOOSING THE RIGHT SCREEN

›› Premade screens come with either a wood or aluminum frame. Aluminum frames are more durable and are not susceptible to warping, as are wood frames after repeated use (making them more expensive). If you are planning to use your screen several times, purchase a screen with an aluminum frame.

›› Screens also come with different mesh counts. Mesh count refers to the tightness in the weave of the screen fabric and represents the number of threads per inch. The higher the count, the smaller the mesh openings will be. Higher mesh counts allow less ink to pass through the screen, and lower mesh counts allow more ink to pass.

WHY IS THIS IMPORTANT? First of all, different materials absorb ink differently. For example, most fabrics are more porous than paper. When working with textiles, you will typically want to use a lower mesh, between 110–160, because the fabric will absorb a lot of ink. If you were to print the same graphic on a poster, as opposed to a T-shirt, you may want to use a higher mesh.

›› Another factor to take into consideration is whether you are printing on a light or dark color. Light-colored materials typically need less ink than darker-colored materials. For example, a black T-shirt might require you to use a lower mesh screen than a white shirt to get the same coverage for the exact same graphic.

The amount of fine detail in your graphics will also help determine the right mesh. To print fine details, use a high mesh count in order to transfer thinner lines. Finally, if you are using specialty inks such as thermochromatic, photochromatic, phosphorescent, conductive, puffy, glitter, and so on, use a lower mesh, as the particles in these inks are large and need a coarse mesh to pass through the screen.

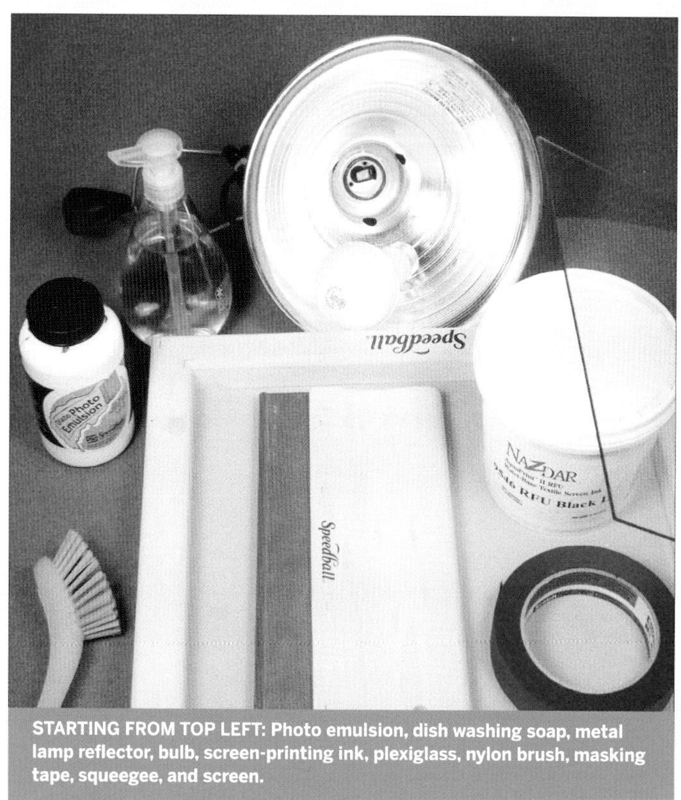

STARTING FROM TOP LEFT: Photo emulsion, dish washing soap, metal lamp reflector, bulb, screen-printing ink, plexiglass, nylon brush, masking tape, squeegee, and screen.

IN SUM **To determine the appropriate mesh count for your screen, you must take into consideration the material you are printing on, the level of detail in your graphic, and the ink that you will be using. For most of your projects, a 110 mesh count is a good starting point.**

CHOOSING THE RIGHT SQUEEGEE

» Squeegees come in a variety of widths, shapes, and blade hardness (often refereed to as the *durometer*). When selecting a squeegee, the width of the squeegee will be determined by the width of the graphic you will be printing. Select a squeegee that's at least a few inches wider than your graphic.

The hardness of the blade determines how much ink passes through the screen. Softer blades deposit more ink than harder blades. In general, the lower your screen mesh, the softer the blade you will want to use. The most common and versatile squeegee is the straight edge. This blade is used for printing on most materials, from T-shirts to posters. A squeegee with a round blade is used for heavy ink deposits or coating applications, typically on fabrics.

SCREEN-PRINTING SAFETY If you are working with oil-based inks and solvents for ink cleanup, follow the safety data sheets and labels found on the containers of the products. Some solvents are highly flammable and toxic and should be stored and disposed of properly. When working with these inks (as opposed to water-based inks), make sure that you work in a properly ventilated area. Always use solvent-resistant gloves for removing inks and emulsion from the screen.

Screen printing

STEP 1: Prepare your workspace.
Screen printing can be a messy process, so it's important to prepare your workspace properly. Before you begin preparing your screen, place newspaper and craft paper around your work area. Also, wear old clothes or an apron to protect your clothing.

STEP 2: Prepare the screen.
Thoroughly scrub your screen with a nylon brush and dishwashing soap. Let the screen dry completely. Using 1" water-resistant masking tape, tape the grooves of the front and back of the screen. Half of the tape (½") should be on the screen, and the other half on the frame. This will prevent ink from leaking from the edges of your screen. Place pushpins, two at each corner, on the back of the screen; this will help keep the screen elevated in order for the emulsion to dry.

STEP 3: Coat the screen.
Using a squeegee, coat the bottom of the screen evenly with a thin layer of photo emulsion. Repeat, evenly coating the inside of the screen. Dry the screen horizontally in a dark, cool area such as a closet, drawer, or inside a cardboard box. You can use a fan to expedite the drying process.

STEP 4: Create the transfer image.
Print a rich, black positive of the image to be screened on a transparency using your darkest print settings. For best results, print two transparency copies of your image and stack them on top of one another. You can use clear tape to align the two copies in place. This will ensure that you have a good transfer without any holes.

STEP 5: Burn the image.
Place the transparencies inside the screen, holding them in place with a transparent piece of glass or plexiglass. Place the screen on top of a dark piece of fabric or paper. Place a light reflector with a high-wattage bulb (a photoflood is recommended) 12" directly above the screen. Burn the

WHAT YOU'LL NEED

- **Premade screen**
- **Screen ink**
- **Photographic emulsion**
- **Emulsion remover**
- **Squeegee**
- **Transparency film**
- **Water-resistant masking tape**
- **Plexiglass**
- **Dishwashing soap**
- **Nylon brush**
- **Cellophane tape**
- **Lamp with metallic reflector**
- **250W–500W photoflood bulb**
- **Dark fabric** larger than your screen
- **Rags**
- **Pushpins**
- **Newspaper or craft paper**
- **Spray bottle**
- **Solvent** to remove oil-based inks
- **Apron** (recommended)

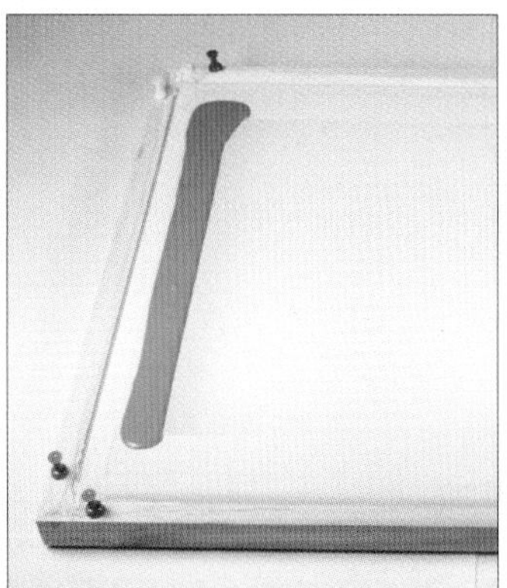

screen for the appropriate exposure time. Refer to the instructions on your photo emulsion or screen-printing kit for the suggested duration.

STEP 6: Remove the emulsion.

Fill a spray bottle with lukewarm water. The water should *not* be hot. Spray the areas of the image with the bottle, opening up the screen. Continue spraying until all unwanted emulsion is removed. Wash and completely dry the screen.

STEP 7: Make the print.

a. Secure the material to be screened (that is, fabric or paper) onto a hard surface. If you are printing onto fabric, it is a good idea to place a piece of cardboard in between the layers of fabric to prevent the ink from bleeding onto the bottom layer (for example, in the center of a T-shirt). Position the screen where you want the graphic to be printed.

b. Slightly lift up the bottom end of the screen. Place the ink horizontally across the end nearest you. Using the squeegee, apply an even smooth coat of ink onto the print area, moving the ink away from you. Lower the screen.

c. Hold the squeegee at a 45° angle and move the ink across the screen back towards you. You should maintain a hard, even pressure on the squeegee during the print stroke.

d. Remove the screen. Using a bottle of all-purpose cleaner for water-based inks or the appropriate solvent for oil-based inks, you can clear out the screen if the screen gets clogged from the ink.

STEP 8: Clean up.

Using warm water and a soft brush, wash all the materials immediately.

NOTE: If you used oil-based inks, you will need to use the appropriate solvent to clean the materials. Make sure you wear solvent-resistant gloves.

The photo emulsion should be stored in a dark, cool place. If you want to reuse the screen, removing the emulsion with photo emulsion remover immediately is strongly recommended; otherwise, you may have a permanent stencil.

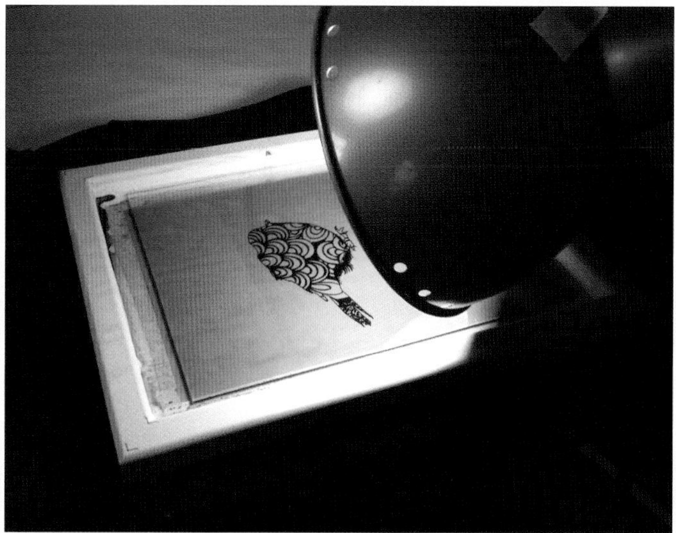

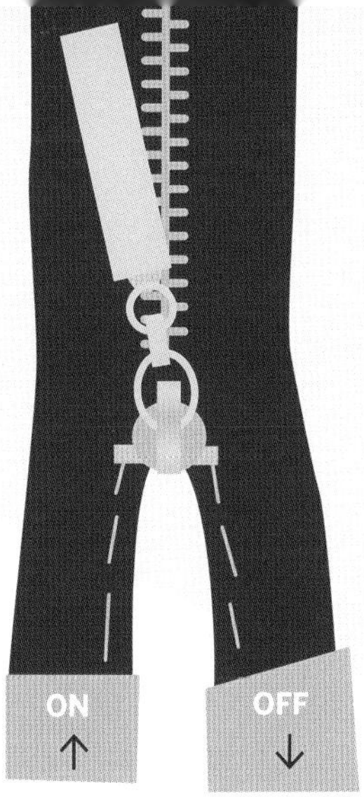

SEWING SOFT CIRCUITS

Electronic textiles are dramatically redefining the way circuits look and feel. The increasing availability of raw conductive materials such as inks, threads, and textiles opens a new world of possibilities to experiment with, to better help you craft electronics into fabric. Circuits can now be hand- or machine-sewn, woven, embroidered, inked, or knit; they can be lightweight, flexible, and even three-dimensional. But don't throw your etchant solution and copper boards away quite yet, as it is difficult and time-consuming to sew complex circuitry by hand. For most projects, you will want to combine traditional printed circuit boards (PCBs) with soft circuits, controls, and switches. The following section introduces basic techniques on how to integrate electronics into textiles to get you started experimenting with soft circuits.

Conductive Threads and Textiles

» Conductive threads and textiles come with varying surface resistivity. Surface resistivity, typically measured in ohms per square, is the resistance of a material to the flow of electric current between opposite sides of its surface. In materials with low electrical resistance, electrons easily flow through or across the surface of the material. Generally, you will want to work with threads and textiles with low surface resistivity.

NOTE: There is a variety of conductive threads and textiles on the market that isn't very conductive, meaning that they have high surface resistivity. Before purchasing any conductive thread or textile, it is important to check the product's specifications to determine its surface resistivity.

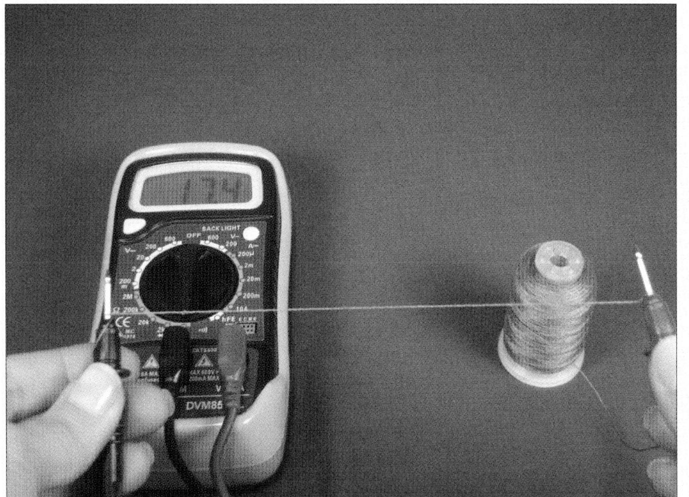

Bookbinder's Knot

» The bookbinder's knot is a great needle-threading technique you can use for hand-stitching conductive paths and sewing electronic components. You will need a needle with an eye large enough to pass the conductive thread through. A needle threader comes in handy if you have difficulty threading the needle.

WHAT YOU'LL NEED

» Needle
» Conductive thread

STEP 1: Snip the end of the conductive thread at a 45° angle to give the thread a sharp edge.

STEP 2: Hold the needle upright and push ½" of thread through the eye of needle.

STEP 3: From the opposite side of the needle, pull 2" of thread through the eye.

STEP 4: Hold the needle in a horizontal position and pierce the tip of the needle through the center of the thread's fibers.

STEP 5: Using your thumb and forefingers, pull the pierced thread toward the eye of the needle.

STEP 6: Pull the other end of the thread taut until the knot is secure.

STEP 7: Knot the bottom of the thread.

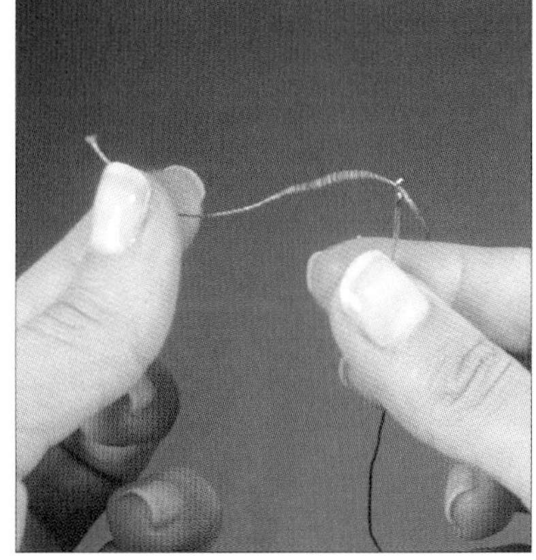

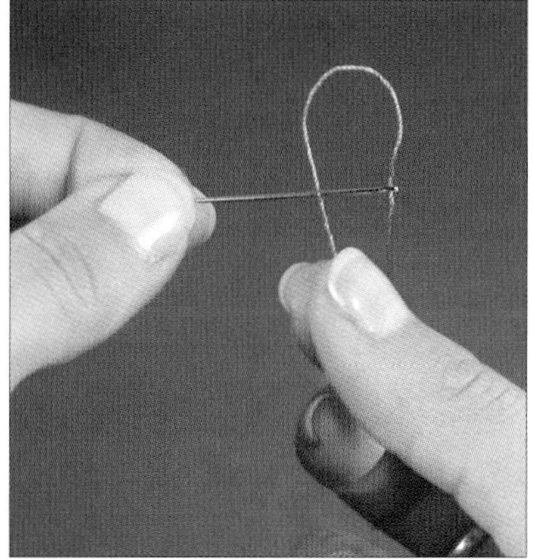

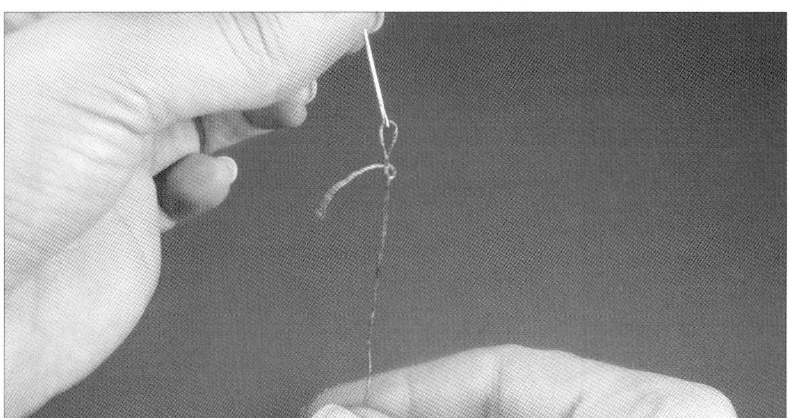

Sewing by Machine: The Perfect Stitch

» When sewing by machine, you will want to use a combination of conductive and regular sewing thread. Unless you're using an industrial or commercial-grade sewing machine, always use the conductive thread as the bobbin thread and regular thread for the top thread. Also, use a needle suitable for medium to heavy-weight fabrics to prevent the needle from breaking.

The perfect machine-sewn conductive path will have the conductive thread isolated on the bottom side of the fabric, while the regular thread locks around the conductive thread and forms a stitch on the opposite side. The conductive thread should not be pulled through to the top of the fabric.

» Unlike traditional wire, conductive thread is not shielded. Any single fray in the thread can cause the circuit to misbehave or even short, especially if several conductive paths are sewn in proximity to one another. To make the perfect stitch, experiment with the different stitch settings on your machine. The results will also vary depending on the fabric you use. Once you have found a setting that works, write it down so that you will have a good starting point for the next project.

At the beginning and end of each conductive path, if both threads are on the same side of the fabric, tie a knot close to the stitch and trim the excess with scissors. You can use a touch of liquid seam sealant to ensure that the conductive thread doesn't fray. If the two threads are on opposite sides, pull the top thread to the bottom of the fabric. Knot, trim, and use a seam sealant to prevent fraying.

If you will be sewing electronic components at the beginning or end of the conductive paths, leave at least 5"–6" of loose thread at the beginning or end of the path. Using the bookbinder's knot, you can use the excess thread to sew the leads of the component directly to the path.

Sewing Components

Electronic components unfortunately aren't manufactured in packages that readily lend themselves to sewing. Fortunately, most components do come with long, pliable leads that can be bent into loops to make hand-sewing possible.

Two tools that will come in handy when sewing electronic components are a sewing needle and a pair of needlenose pliers.

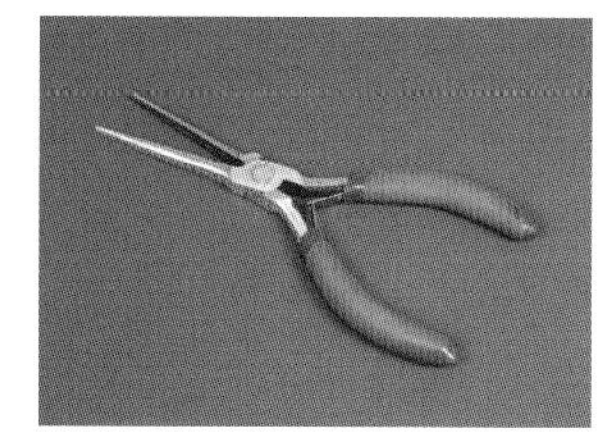

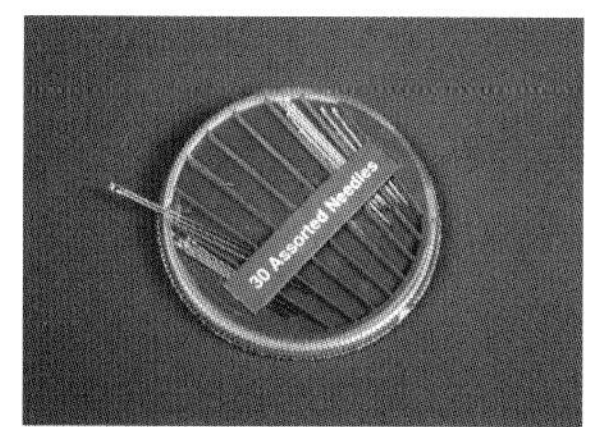

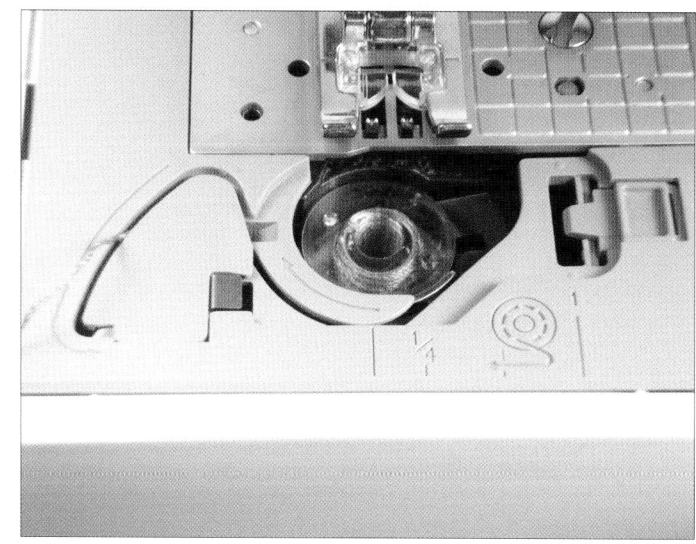

Sewing Components with Long Leads (LEDs, Resistors, Capacitors)

WHAT YOU'LL NEED

» Needle
» Conductive thread
» Component with long lead
» Fabric
» Liquid seam sealant

STEP 1: Using a sewing needle, pierce the fabric at the location where you want to place the leads.

STEP 2: Slip the leads of the component through the pierced holes to the opposite side of the fabric.

STEP 3: Using needlenose pliers, gently twist the leads around the tip of the pliers a few times, creating loops. Using the needlenose pliers, gently bend the loop flush to the fabric.

STEP 4: Using a needle with conductive thread, stitch around the loop several times, securing the loop to the fabric.

STEP 5: Once all the electronic components are sewn into the circuits and you have ensured that the circuit is working properly, you can add a touch of a liquid seam sealant to each component lead and over each sewn conductive path to ensure that the conductive thread doesn't fray over time.

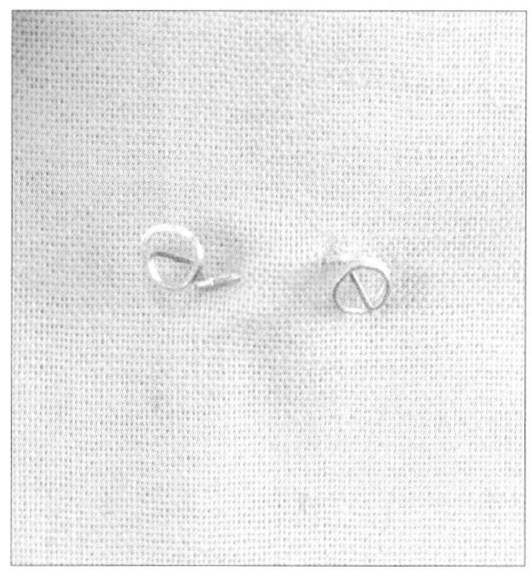

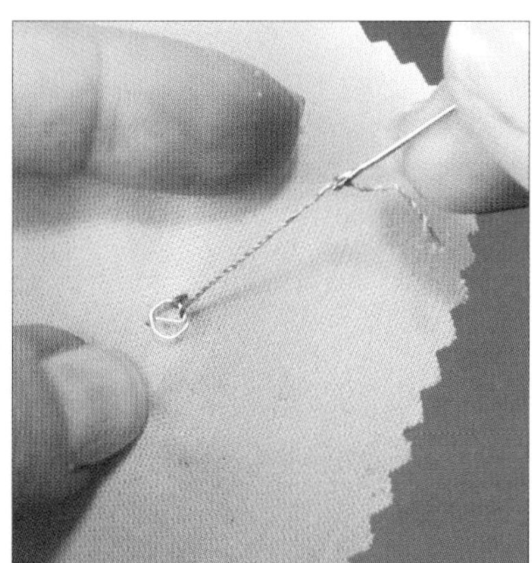

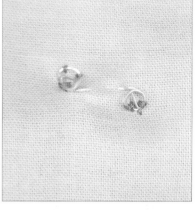

TUTORIAL

Sewing Integrated Circuit (IC) Chips

» When sewing ICs, use a *DIP (dual inline package) socket*, a connector designed to hold IC chips. They are inexpensive and make it easier to address each individual lead without risking damaging the chip. Look for DIP sockets with long leads to make sewing easier.

WHAT YOU'LL NEED

» Needle
» Conductive thread
» Dip socket
» Fabric
» Liquid seam sealant

STEP 1: Gently push the leads of the DIP socket through the fabric.

STEP 2: Using needlenose pliers, bend the leads flush to the fabric.

STEP 3: Using a needle with conductive thread, stitch around the first lead several times. Continue, and make a straight stitch about 2" in length. Repeat for all the leads, making sure that the conductive thread from each lead does not intersect or touch the conductive thread from the lead beside it.

STEP 4: Once all the leads are sewn, you can add a touch of a liquid seam sealant to each lead and accompanying conductive path, to ensure that the conductive thread doesn't fray over time.

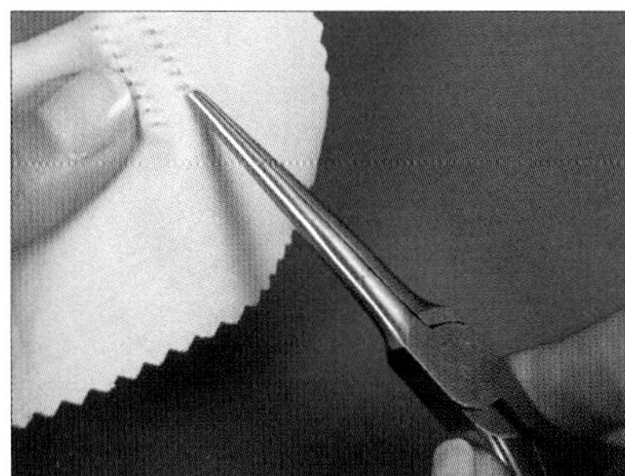

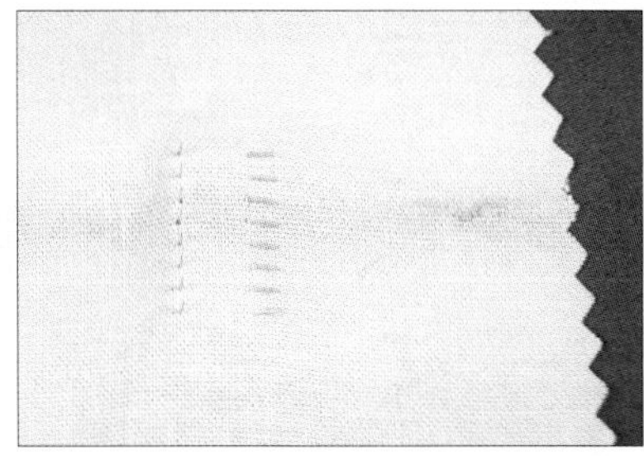

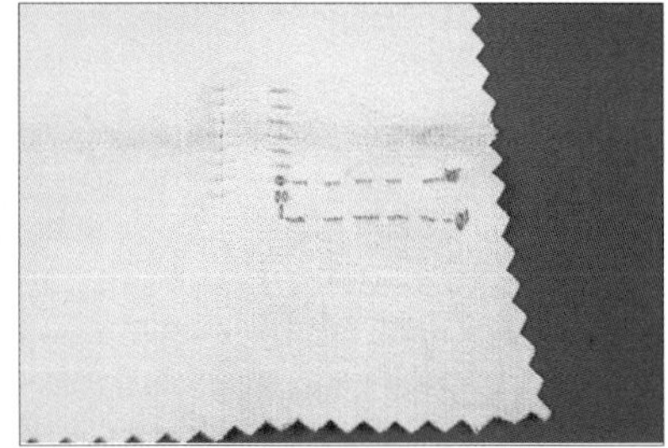

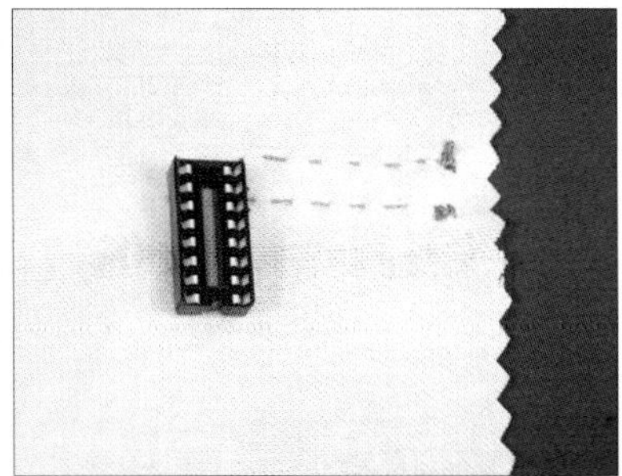

TUTORIAL

Sewing Components with Wires

» For components with wires (battery holders, piezo speakers, motors, and so on), you will first have to solder metallic loops to the end of each wire in order to sew them in place. You can use ordinary nickel-plated rounded or infinity-shaped eyes or similar metallic loops used in beading and jewelry making.

WHAT YOU'LL NEED

- Needle
- Conductive thread
- Component with wires
- Fabric
- Sewing eye fastener

STEP 1: Strip about a ¼" of insulation from each wire. If the wire is stranded, using your thumb and forefinger, twist the strands together.

STEP 2: Loop the stripped end of each wire around the metallic loop.

STEP 3: Using a soldering iron, add a touch of solder.

STEP 4: Using a needle with conductive thread, stitch around the loop several times, securing it in place.

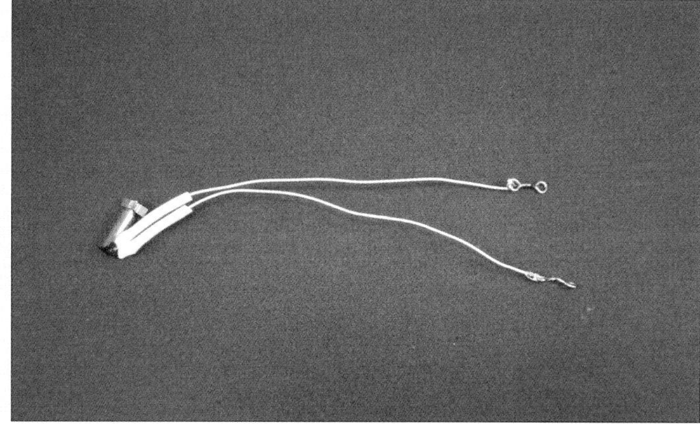

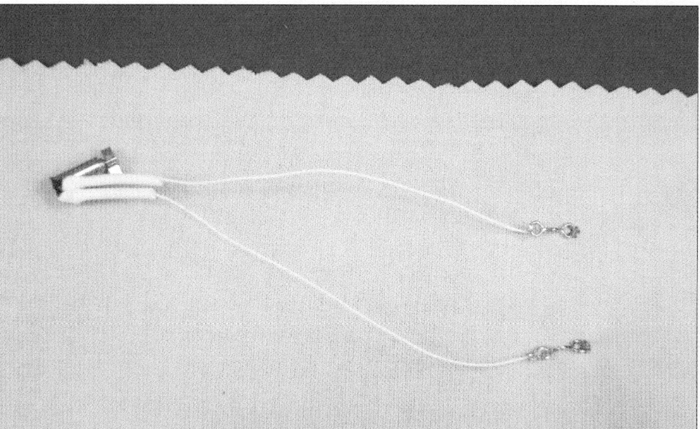

Making Switches from Fastening Devices

Following are a few examples of simple switches made from different snaps, hook and eyes, zippers, and grommets.

A switch is a device that is used to open (disconnect) and close (connect) circuits by mechanical or electronic means. When working with soft circuits, traditional electronic switches can be easily substituted for ordinary fastening devices such as metallic snaps, grommets, and zippers typically used in garment construction and jewelry making.

SWITCHES FROM SNAPS

Metallic snaps make ideal switches, as they are composed of two parts: a female and a male counterpart, which interlock when pressed together. To make a switch using snaps, you need an opening or break in the conductive path of the circuit. Ideally, you would have the opening near the power source. At the end of each open path, you can sew the female and male ends of the snap. When the snaps interlock, the circuit is closed, allowing electricity to flow through the circuit.

SWITCHES FROM METALLIC ZIPPERS

With the help of conductive thread or ink, zippers can be transformed into control devices for electronically enhanced wearables and accessories. To turn a zipper into a switch, the upper or lower two teeth opposite from each other can be sewn with conductive thread. A conductive path from each zipper tooth can be continued to the rest of the circuit. When the zipper is closed or opened with the slider, the two opposing teeth connect, closing the circuit. Each set of teeth on opposing rows of the zipper can also be individually addressed to create a switch. The main drawback of using conductive thread (or ink as an alternative) in the middle of the zipper is the eventual wear and tear on the thread (or ink) from the slider.

SWITCHES FROM MAGNETS

Magnets make excellent switches (not to mention clasps) for haute-tech accessories. Similar to snaps, to make a switch using magnets, you need an opening in the conductive path of the circuit. At the end of each open path, you need to secure the magnet using conductive epoxy, thread, or fabric. When securing them in place, keep in mind that the two magnets on opposite ends should attract (not repel) one another.

SWITCHES FROM GROMMETS

Another creative switch you can incorporate into wearables is a drawstring switch. The basic components needed to make a drawstring switch are grommets, conductive fabric, and a drawstring.

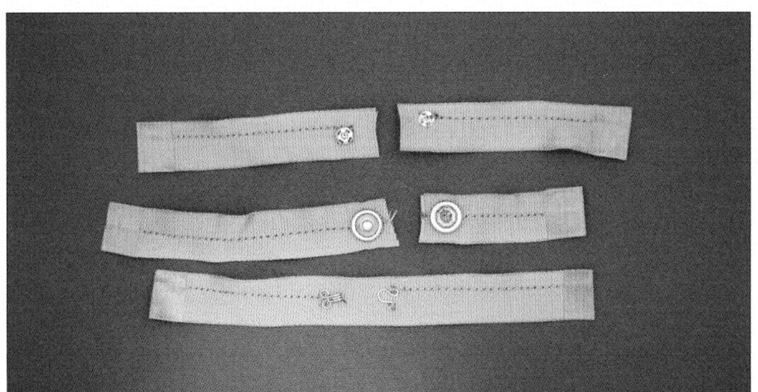

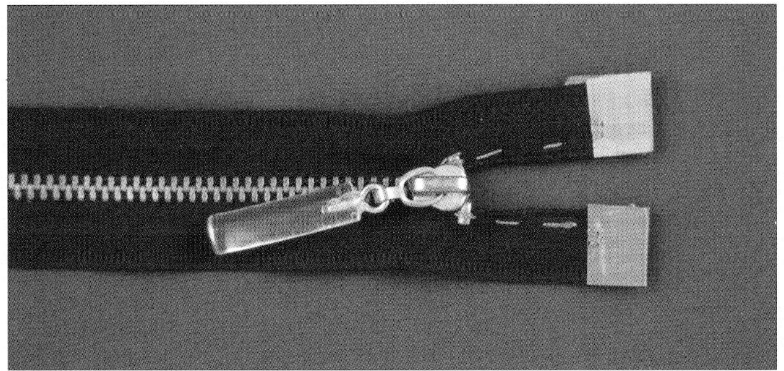

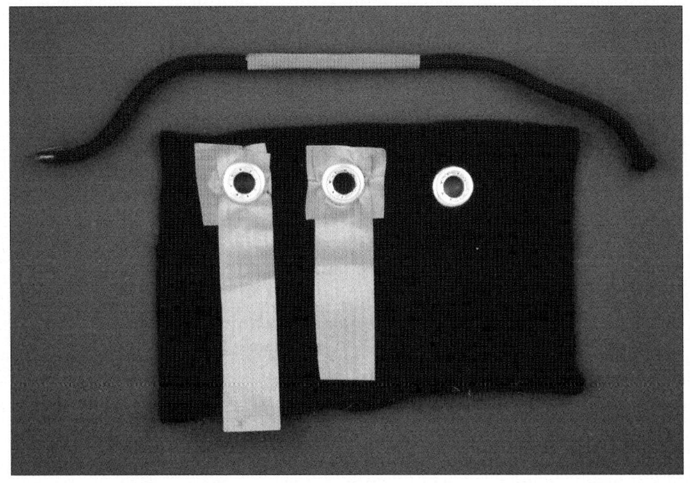

Making Soft Switches (or Pressure Sensors)

» **To make a soft switch, all you need are two layers of conductive material (fabric or a patch of thread sewn onto fabric) that are slightly separated from each other.** You can use a separate piece of thicker fabric with the center portion cut out or a piece of netted fabric, such as tulle, in the middle of the two conductive layers.

» When you apply pressure to the top conductive layer, the circuit is closed, allowing electricity to flow. By changing the thickness of the middle fabric layer, you can adjust the sensitivity or the amount of weight and pressure it takes to close the circuit.

» The primary advantage of soft switches over traditional switches is that they can be designed to easily conform to a 3D surface, such as your body.

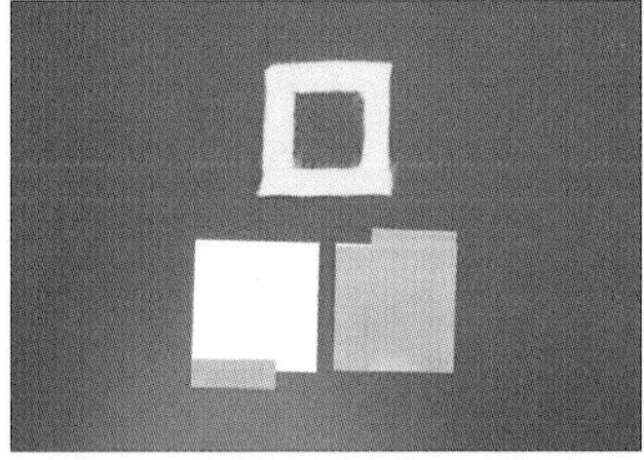

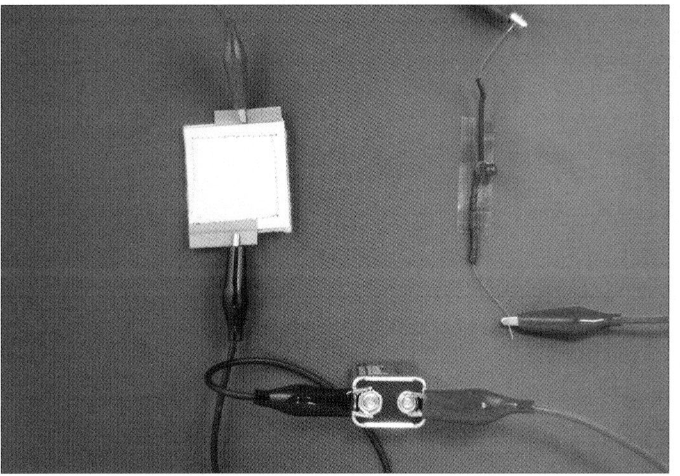

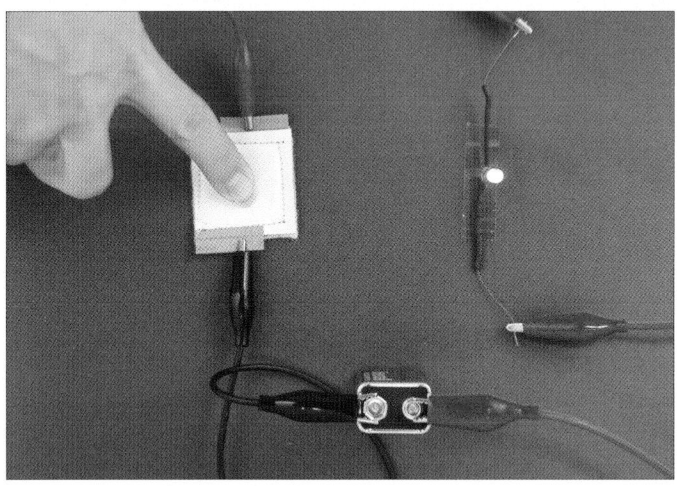

TUTORIAL

Making Resistors

» To make soft resistors, experiment using different stitch patterns on your sewing machine. Using a multimeter, test the resistance of each stitch. You can either stitch the resistor directly into your circuit path or create individual soft components.

WHAT YOU'LL NEED

- » **Magnetic jewelry clasp** metallic
- » **Super stong magnet**
- » **2"×6" piece of fabric**
- » **1"×5" piece of fabric**
- » **Sewing eye fasteners (2)**
- » **Conductive thread** with high resistivity
- » **Conductive thread** with low resistivity
- » **1" conductive fabric tape**
- » **Sewing machine and sewing needle**
- » **Alligator clips (3), LED, and 9V battery**
- » **Multimeter**

There are many ways to create variable resistors. The following tutorial is an example of how to create a variable resistor using two magnets.

In this tutorial, a piece of fabric is sewn with highly resistive conductive thread and with two conductive contacts placed on opposite sides. The resistive fabric will have low resistance at one end that will gradually increase to a maximum resistance on the opposite end. One contact will connect to the power source while the other will connect to ground of the circuit. Two magnets, placed on opposite sides of the fabric, are manually moved across the resistive fabric layer, completing the connection.

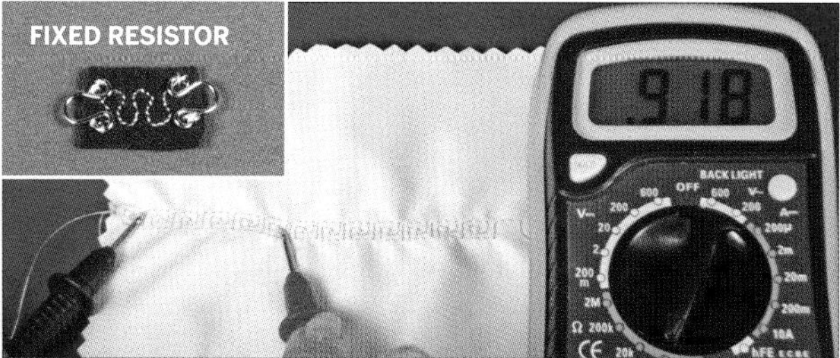

STEP 1: Cut one piece of fabric to at least 2"×6" in size. Place a bobbin of conductive thread with high resistivity in the machine. Starting from the left edge, machine-stitch a 5" pattern, leaving a minimum of 1" not sewn on the opposite side.

STEP 2: Set your multimeter to measure resistance. The resistance function is usually marked on the meter by the symbol Ω. The resistance value should increase as you move further along (to the right of) the conductive path.

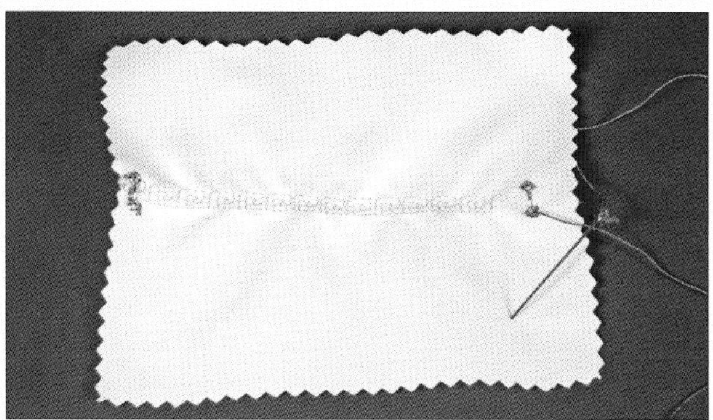

STEP 3: Using the conductive thread with low resistivity, sew an eye directly onto the conductive path on the left side. Knot and cut the thread. Sew another eye, on the opposite side, about ½" away from the conductive path. Do NOT cut the thread at this point.

STEP 4: Grab the magnetic jewelry clasp. Cut a piece of conductive fabric tape slightly larger than the diameter of the clasp. Place it on the bottom of the magnet and wrap the edges of the conductive tape around the clasp. This will ensure that the metal housing of the clasp and the magnet have an electrical connection.

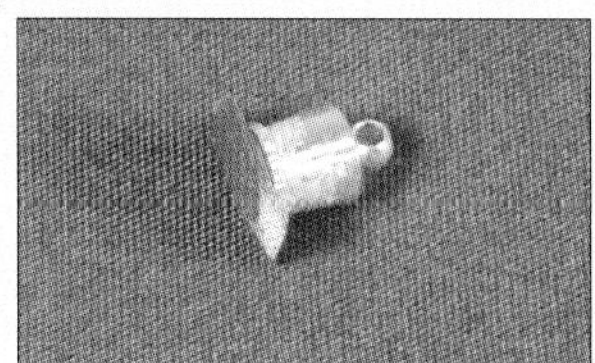

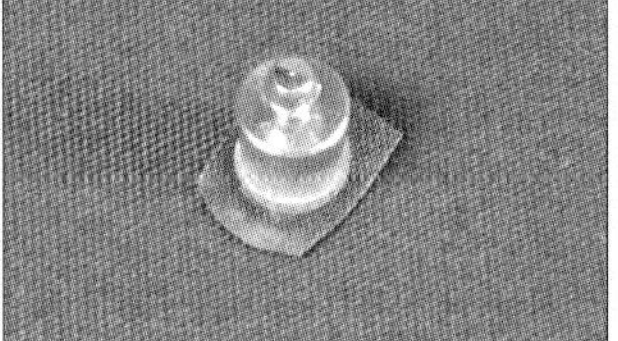

STEP 5: Leaving 5" of loose thread, thread the needle through the loop of the magnetic clasp. Knot and cut the thread.

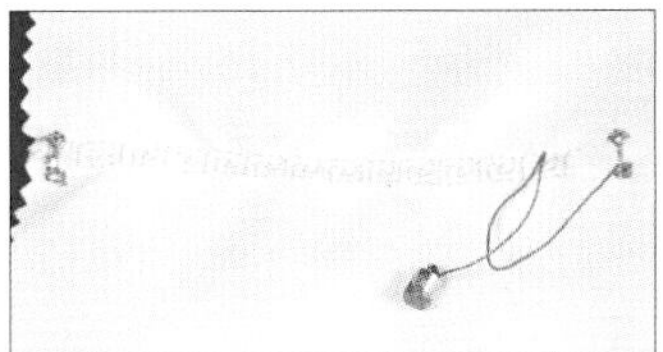

STEP 6: Flip the fabric over so it is wrong side up. Cut a piece of fabric ¼" wider and longer than the width and length of the sewn conductive path. Grab a magnet and the second piece of fabric. The second piece of fabric will be used to create a pocket around the magnet. Place the magnet directly on top of the sewn conductive path. You want to make sure that you place the magnet in the right direction so that will attract and not repel the magnetic clasp on the opposite side of the fabric. Center the second piece of fabric directly on top of the first, covering the magnet. Using ordinary thread, machine or hand-sew the fabric around all edges. The magnet, sandwiched between the two pieces of fabric, should have ample room to move around. Flip the fabric over and place the magnetic clasp directly over the second magnet sandwiched in between the fabric layers. You should be able to move both magnets along the conductive path.

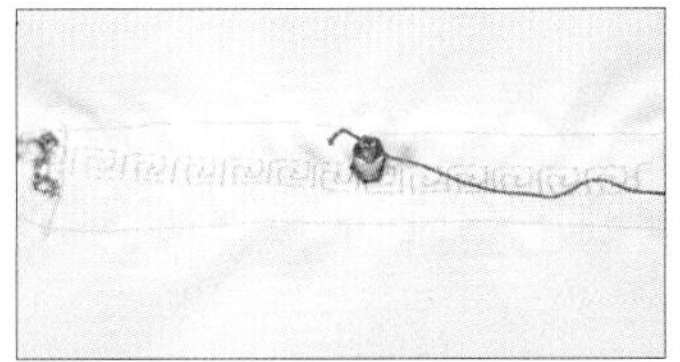

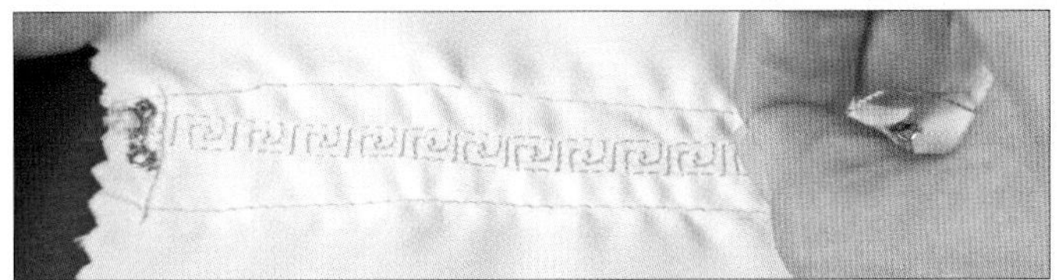

STEP 7: Test the variable resistor using three alligator clips, an LED, and a 9V battery. Connect one end of an alligator clip to the eye on the left side of the fabric and the opposite end to the negative battery terminal. Connect one end of another alligator clip to the positive battery terminal and the opposite end to the positive lead of the LED. Connect one end of the last alligator clip to the negative lead of the LED and the opposite end to the eye on the right side of the fabric. As you move the magnetic clasp from left to right, the LED should get dimmer in brightness.

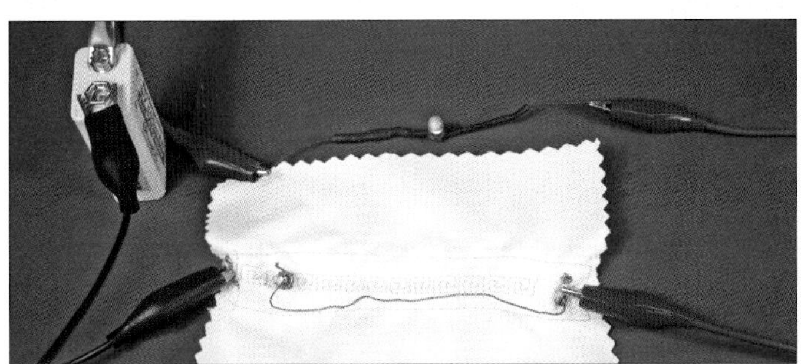

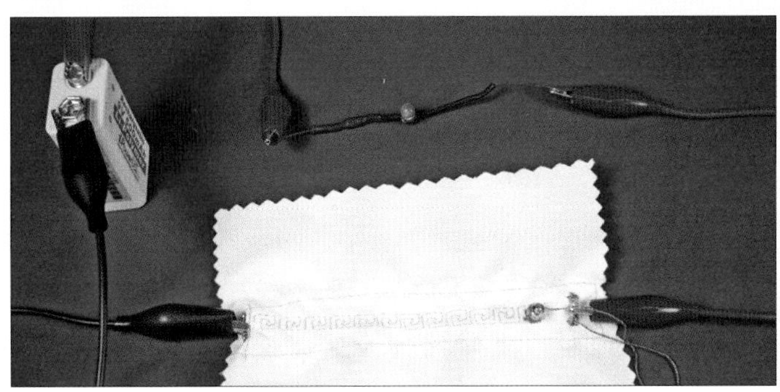

TROUBLE-SHOOTING CIRCUITS

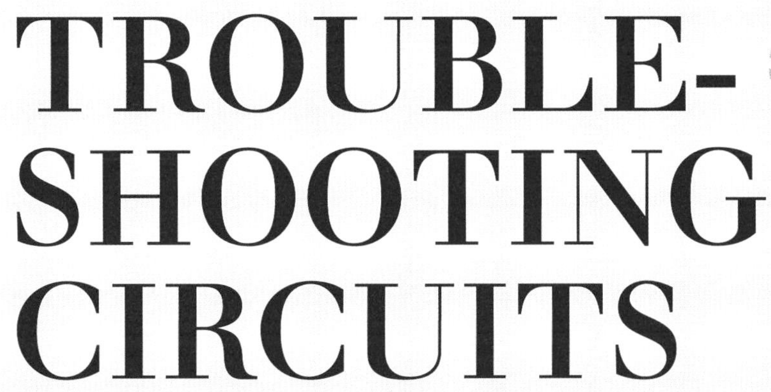

Half the fun of building circuits is troubleshooting them. You will almost certainly encounter some glitch that will require you to spend a good amount of time trying to unravel exactly what went wrong and where. This detective work is a large part of the process (especially the learning process) and part of the challenge in working with electronics. Even if you have done everything perfectly, you may have a faulty component in the mix that will cause the circuit to misbehave or simply to not work at all.

Armed with the right tools, patience, and know-how, you can learn to detect and solve problems quickly. Remember, every circuit that you troubleshoot teaches you something.

Check Your Connections

» The first rule of thumb in troubleshooting circuits is to check the connections. Follow the circuit path from power to ground, making sure that you have connected every component properly. Once you're certain that the components are wired properly, you can continue to troubleshoot using a multimeter.

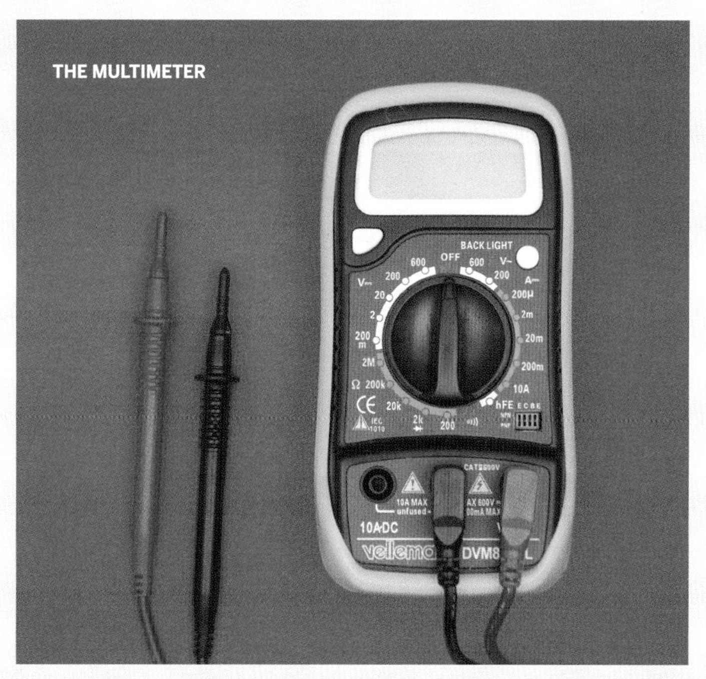

THE MULTIMETER

Using a Multimeter

A multimeter is an indispensable device that will tell you whether you have a weak connection, continuity between two components, enough juice to get your circuit running, and much more.

» There are two types of multimeters: *digital* and *analog*. Digital multimeters are recommended because they are easier to work with. Digital meters can be normal or autoranging meters. The autoranging multimeters automatically detect the range of the component and give you a value, whereas normal meters require you to set the range first manually before they can calculate a value. As meters produced by different manufacturers vary in design, you should read your product's manual to become familiar with that specific multimeter.

Digital multimeters generally have an LCD to show calculated values, a dial switch to select what you want to measure (current, voltage, resistance, and so on), a number of jacks, and two colored test probes that plug into the jacks. The two colored test probes are typically red and black. Generally, the black probe plugs into the negative ("–") or "common" jack, marked by "com." The red probe needs to be plugged into the jack associated with a particular type of measurement (for example, current, resistance, or voltage). Because multimeter designs differ, you must refer to your manual to determine the proper jack for each probe.

» The first step in troubleshooting any circuit is checking the voltage between power and ground.

TUTORIAL

Measuring the Voltage of a Circuit

WHAT YOU'LL NEED

- » Battery
- » Battery holder
- » Resistor
- » LED
- » Multimeter
- » Alligator clips (3)

STEP 1: Connect the power supply to the circuit.

STEP 2: Connect the red probe to its appropriate jack (refer to your multimeter's manual) and select the appropriate range for the voltage being measured. For example, if you are using a 9V battery to power the circuit, you need to select the closest range higher than 9V. It's important to distinguish between DC volts and AC volts. Because you will be working only with DC volts, you should select the DC voltage option marked with a solid line over a dashed line.

STEP 3: Touch the black probe of the multimeter at a ground connection of the circuit (normally the negative terminal of the battery). You can use an alligator clip to connect the black probe to the negative terminal of the battery temporarily.

STEP 4: Touch the red probe where you want the voltage to be measured. The voltage across the entire circuit should be near the value of the power source. If the value is negative, switch the probes to get the right polarity.

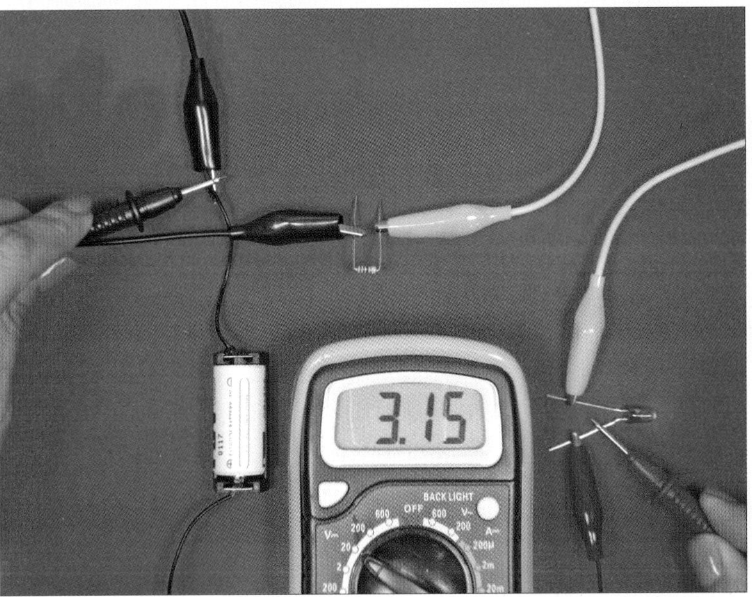

NOTE: The voltage setting also comes in handy when you need to determine the polarity of a battery, solar cell, or the wires of a power adapter. Using the red (positive) and black (negative) probes of your multimeter, hook the probes randomly to the contacts or wires to see if the voltage measured is negative or positive. If the reading is negative, reverse the leads. Now you know that the contact/wire that the black lead is attached to is negative (ground), and the other is the positive.

Measuring the Current in a Circuit

WHAT YOU'LL NEED

» 9V battery
» Alligator clips (2)
» 100K resistor

STEP 1: Connect the power supply to the resistor using the alligator clips.

STEP 2: Plug the red probe into the correct socket for the measurement to be made. If you are unsure of the current measurement, plug the socket into the one marked 10A.

STEP 3: Switch the multimeter to measure current, selecting the closest range for the current being measured.

STEP 4: Temporarily disconnect the battery from the circuit and open the circuit between the resistor and the positive battery terminal. Connect the positive red probe to the alligator clip connected to the positive battery terminal and the black negative probe to the alligator clip connected to the resistor. Connect the battery to the circuit again.

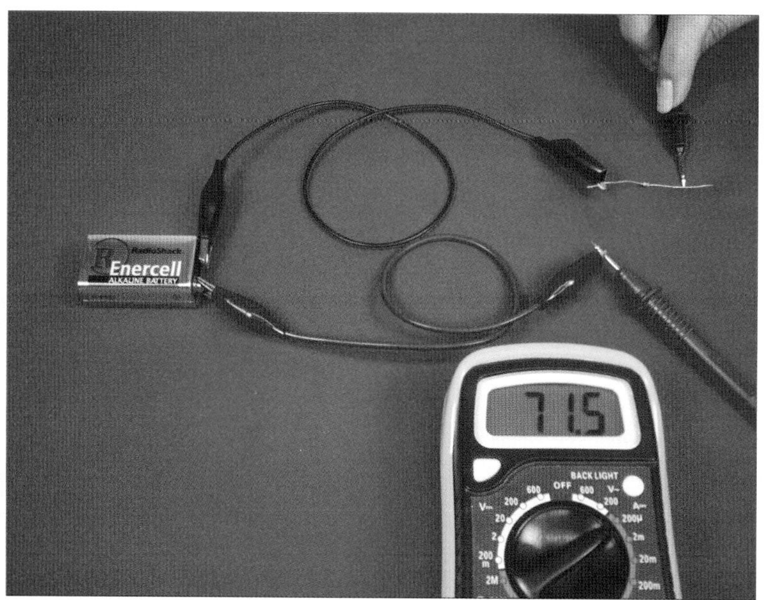

NOTE: To measure the current of a more complex circuit, you need to make sure that the multimeter is connected to the circuit in series so that all the current flows through the meter.

Measuring Continuity

» Checking for continuity in a circuit allows you to test whether two points are electrically connected.

WHAT YOU'LL NEED

» Circuit board
» Multimeter

This is important if you want to ensure that points of a circuit that you think are connected really are connected and that the solder joints and sewed connection points are working effectively. Most meters have an audible beep or tone to verify that continuity exists between two points in a circuit.

⚠ **Make sure to disconnect your power source before testing.**

STEP 1: Set the meter to the continuity setting (refer to the product manual). Typically, the continuity setting on a meter resembles a speaker or Ω for resistance.

STEP 2: Touch the two ends of the probe at the points where you want to test whether continuity exists between two points in the circuit. During a continuity test, the probes are interchangeable — you don't have to worry about the polarity of a circuit. If continuity exists, the meter should beep. If the meter does not beep, check the connections.

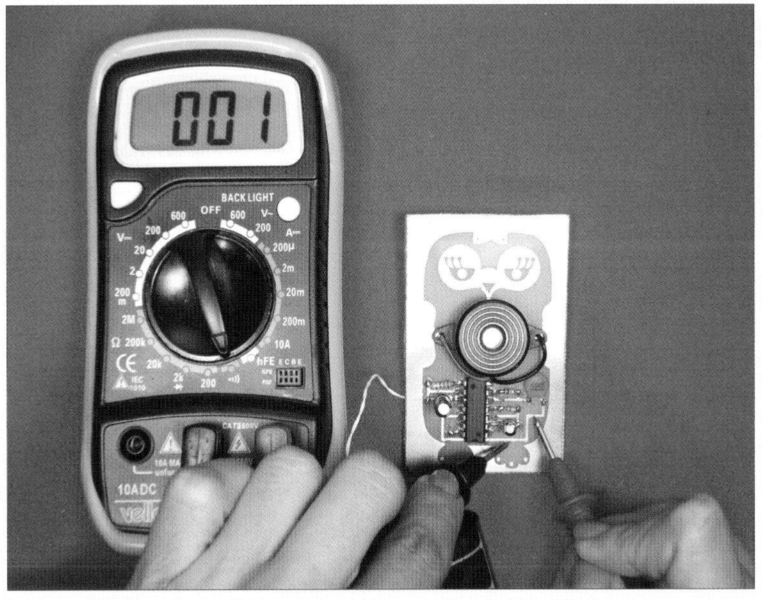

Some meters will display "OL" (meaning open loop) to signify that there is no continuity, while others display a very large number (indicating a high resistance). In these particular meters, a low number is displayed (typically under 100 Ω) to indicate continuity. Ideally, you want a reading as close to zero as possible.

Determining the Polarity of an LED

LEDs are generally manufactured with a number of visual cues that will help you determine polarity if you don't have the manufacturers datasheet. With the help of a multimeter or a simple coin cell battery, there are also a number of quick electrical methods to determine polarity.

FOR STANDARD LEDS

» The longer lead is the anode (+) and the shorter lead is the cathode (-).

» The LED case exterior is curved on the anode side and flat on the cathode.

» If you look inside the plastic casing of an LED, you will notice one lead is wider than the other. The narrower lead is typically the anode and the wider lead is the cathode.

» Use a 3V flat lithium cell to test the LED by touching the LED across the battery terminals. When the LED lights, you will know you have the positive lead of the LED touching the positive side of the battery.

FOR SMD LEDS

» The back of surface mount (SMD) LEDs have either a green arrow or line indicating the negative contact.

» Some multimeters output enough voltage to test LEDs using the Diode Check setting. Some meters will beep and/or show the forward voltage, and/or the LED will light up when connected in the correct polarity (anode +, cathode -).

FOR PIRANHA OR HIGH-FLUX LEDS

» Create a simple circuit using alligator clips, a resistor, and a battery. Refer to page 29.

FOR BI-COLOR OR TRI-COLOR LEDS

» The center is the common cathode (-).

» The outer leads are the anodes (+), one for each color.

NOTE: Bi-color and tri-color LEDs do not follow the conventional long lead is anode (+) rule.

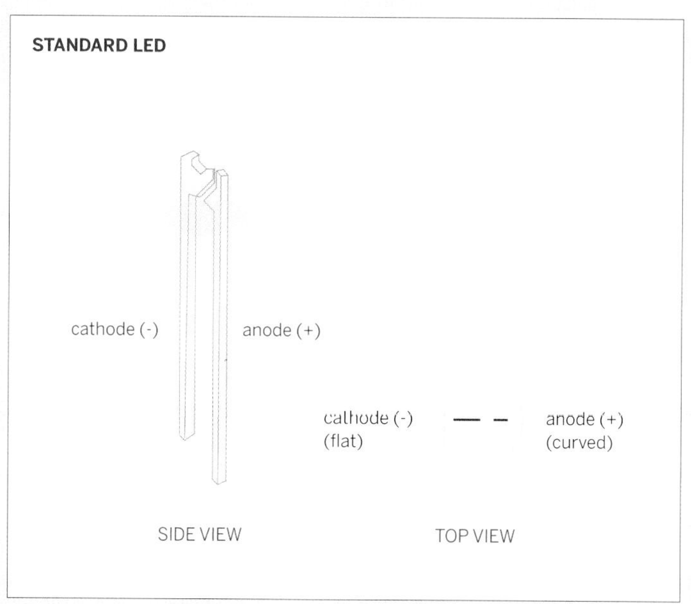

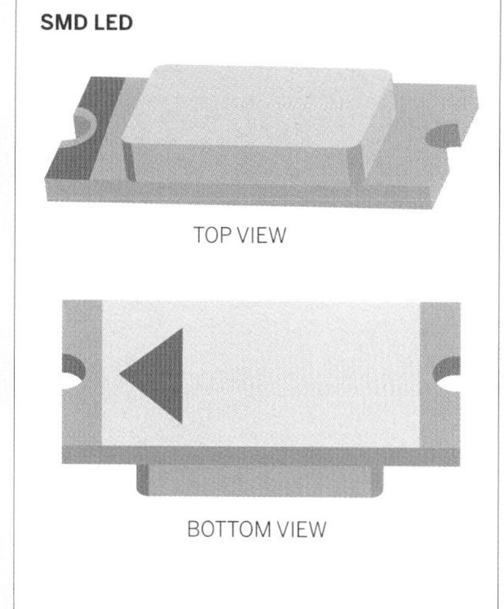

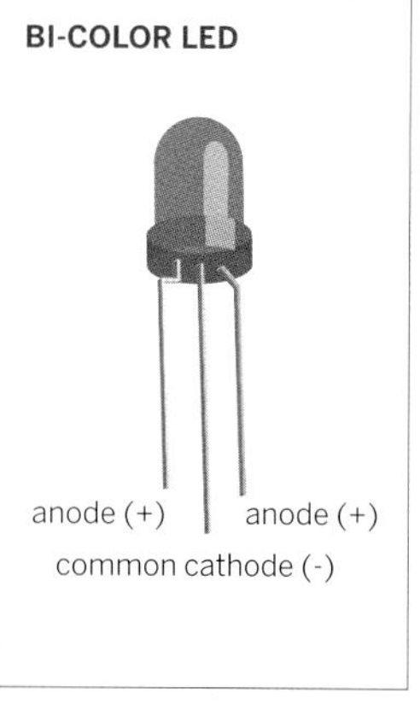

Measuring Resistance

WHAT YOU'LL NEED

» Multimeter
» Resistor

» Multimeters are also used to measure the resistance value of a resistor.

To measure the value of a resistor, follow these three steps:

STEP 1: Set the meter range greater than the resistance being measured. If you are unsure of the resistance value, select the highest setting available. The resistance function is usually marked on the meter by the symbol Ω.

STEP 2: Remove the component from the circuit.

STEP 3: Touch the red probe to one lead of the resistor and the black probe to the other. If the meter reading is close to zero, select a lower resistance range until you get a readable value.

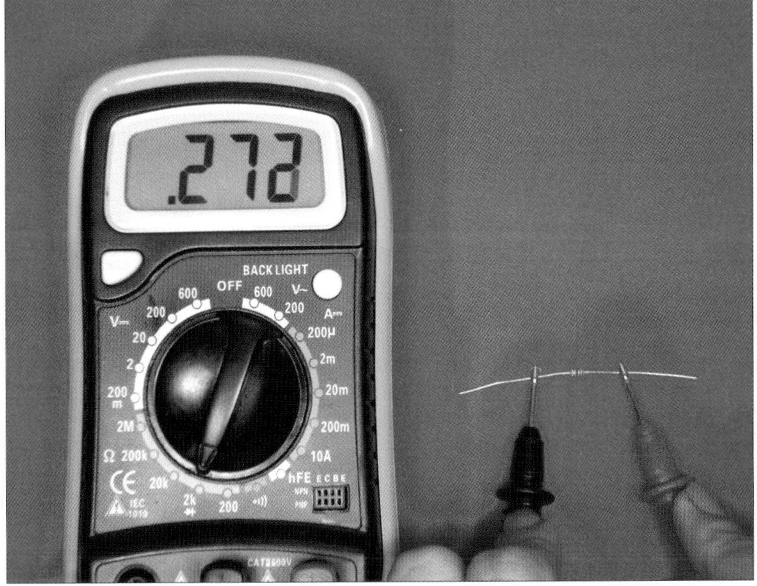

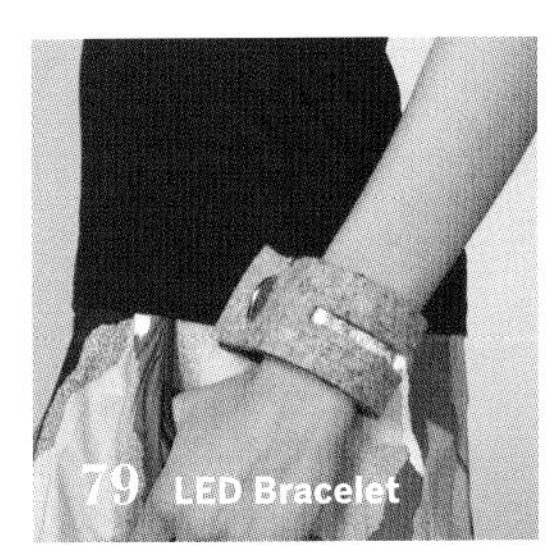

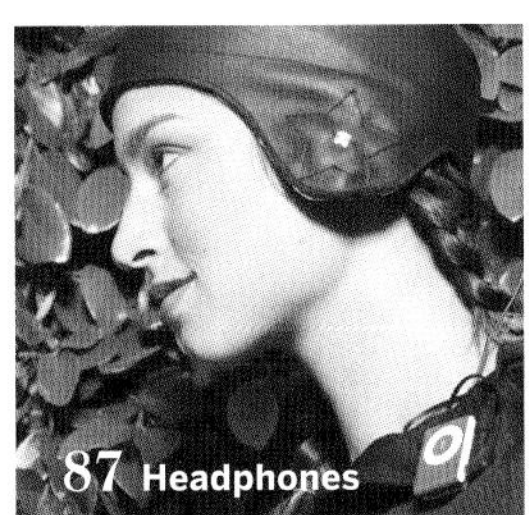

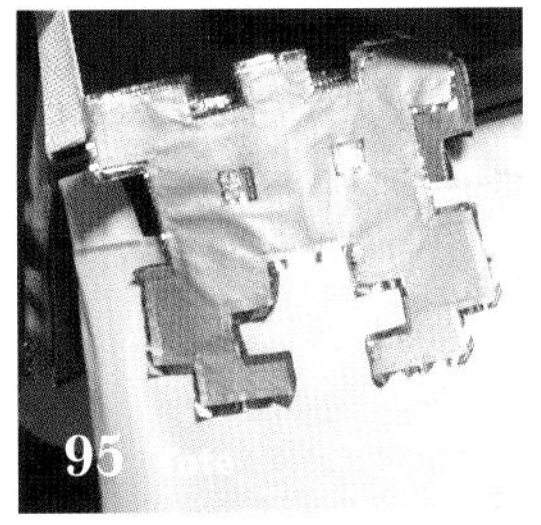

// **The future of crafting fashion has finally arrived! Rockers, crafters, fashionistas, and techno geeks, get ready to update your wardrobe with high-tech textiles and twinkling LEDs.**

WEARABLES

Now you can saunter into the night dressed in light! Look smart and crafty with a handmade tote that twinkles when you receive a cellphone call. Or get stylishly geared and rock out to your favorite tunes with fuzzy headphones that keep you toasty and noticed on a brisk morning jog. When fashion, technology, and craft merge, you get custom clothing and accessories embellished with techno-wizardry that illuminates your personal style and expressive flair. This chapter features four tutorials to get you starting in creating your very own techno-atelier.

LED BRACELET

Who needs diamonds when you have LEDs! Add a little sparkle to your little black dress with a bespoke LED Bracelet cuff. Simple and streamlined, this LED cuff is made from raw industrial felt and inlaid high-flux LEDs. It's the perfect accessory. Donning this haute-tech cuff will certainly make you the light of the party.

HOW IT WORKS »

The LED Bracelet incorporates three square LEDs inlaid into industrial felt. The coin cell battery in this simple circuit is used both as a mechanical clasp to hold the two sides of the bracelet together and as a switch to power and turn the bracelet on when it is worn. When the cuff is worn and the battery is slipped into its slit, the LEDs turn on. When the cuff is not in use, the battery is stored in the pocket of the inner lining of the bracelet.

COMPLEXITY: BASIC

WHAT YOU WILL LEARN

- » Work with LEDs
- » Make soft battery holder

RELATED TUTORIALS

- » Working with LEDs (19-23)
- » Prototyping a Simple Circuit Using Alligator Clips (29)
- » Determining the Polarity of an LED (74)
- » Sewing Soft Circuits (57-67)

MATERIALS

- » **Industrial felt**
- » **Decorative fabric**
- » **Solid-colored fabric**
- » **Conductive thread**
- » **Cotter pins (2)**, 1¼"–1½", found in any hardware store
- » **Magnets (4)**, ½" diameter, strong
- » **Colored thread**
- » **Heat-shrink tubing, 4"**

TOOLS

- » Scissors
- » Sewing needle
- » Needlenose pliers
- » Pencil or marking pen
- » Hot glue gun
- » Utility knife
- » Sewing machine
- » Heat gun or hair dryer

ELECTRONICS

- » Piranha high-flux LEDs (3)
- » 3V 20mm–25mm coin cell battery

TEMPLATES

- » Template A
  Industrial felt top layer (1)
- » Template B
  Bracelet back lining (1)
- » Template C
  Battery pocket (1)
- » Template D
  Decorative fabric (middle) layer (1)

TEMPLATES

Templates available online at fashioningtechnology.com/ledbracelet.

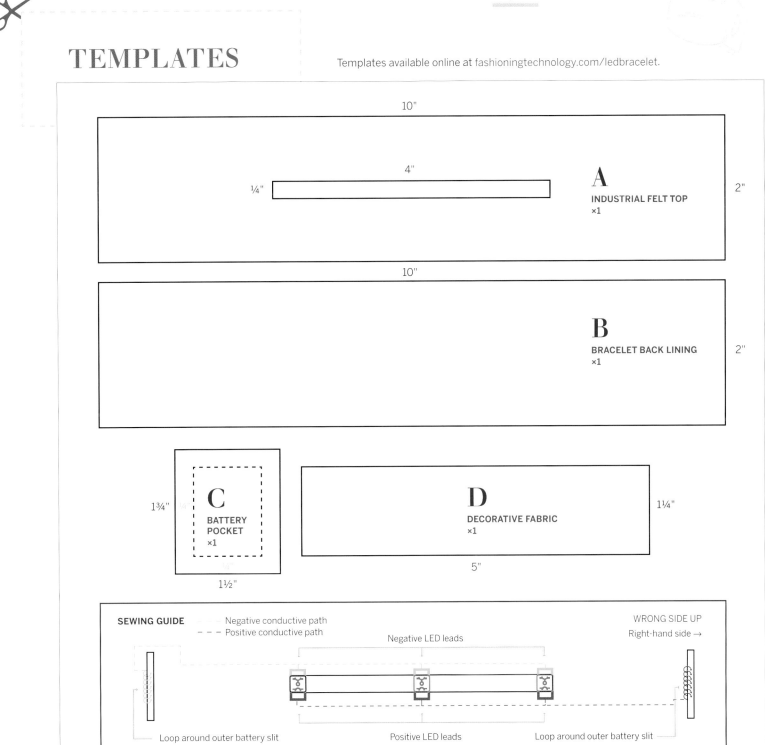

10"

4"

¼"

A

INDUSTRIAL FELT TOP
×1

2"

10"

B

BRACELET BACK LINING
×1

2"

1¾"

¼"

C

BATTERY POCKET
×1

¼"

1½"

D

DECORATIVE FABRIC
×1

1¼"

5"

SEWING GUIDE

Negative conductive path
Positive conductive path

WRONG SIDE UP
Right-hand side →

Negative LED leads

Loop around outer battery slit

Positive LED leads

Loop around outer battery slit

START »

1. CUT THE TEMPLATE

a. Using a marking pen, trace Template A onto the industrial felt. Using a utility knife, cut out the template.

b. To determine the length of the bracelet and the battery slit location for your wrist, wrap the industrial felt around your wrist loosely, making sure to align the edges of both sides. Pinch the opposite sides of the felt close to your wrist until it feels snug. Using a marking pen, mark the location.

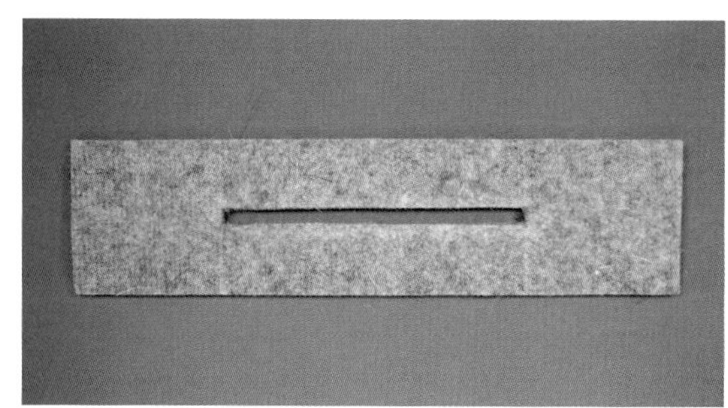

c. Using a marking pen, make another mark about 1" from your original mark; this is the location of your battery slit. Using a utility knife, make a vertical slit slightly smaller than your coin cell battery. Repeat, creating a slit on the opposite end of the felt. You should now have 2 battery slits (on the left and right sides) that align when you wrap the felt around your wrist.

d. Wrap the felt around your wrist until both edges align. Slip the battery through the battery slits, making sure the bracelet fits comfortably. If you have a small wrist, you may want to cut off the extra material from the edges of the bracelet. Leave at least 1" of material on both edges. Make adjustments to Template B accordingly.

Using a marking pen, trace Template B and C on the colored fabric and Template D on the decorative fabric. Using scissors, cut out all 3 templates.

2. SEW THE CIRCUIT PATH

a. Place the industrial felt layer wrong side up. Using conductive thread, loop the thread several times around the *outer* battery slit on the *left* side of the felt. This will be the negative contact for the battery. Continue sewing a path to the top of the felt until you reach the edge of the rectangle cutout.

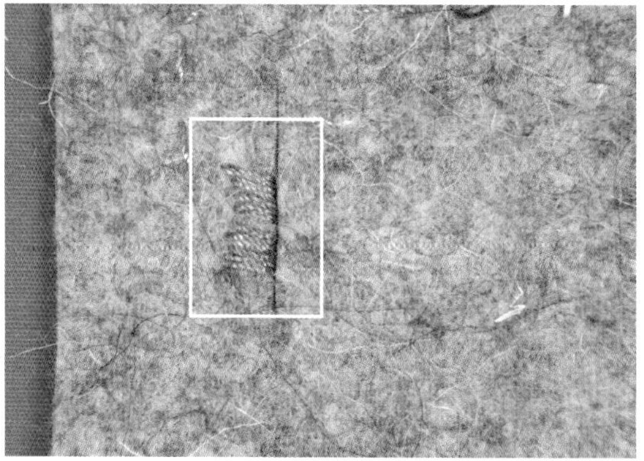

b. The high-flux LEDs have 4 legs: 2 positive (anodes) and 2 negative (cathodes). The best way to distinguish the positive and negative leads is to refer to the datasheet or wire them temporarily using alligator clips (see page 29).

c. Once you have determined the positive and negative leads, position the LEDs wrong side up inside the rectangle cutout with *all* the negative leads facing up and the positive leads facing down. You will be sewing the LEDs together in parallel (see page 22). Using needlenose pliers, carefully bend the leads flush to the fabric.

d. Continue the path from the negative battery slit to the LEDs, securely sewing the negative leads to the felt. *Sew only along the surface of the felt and not through it to avoid having the conductive thread visible on the opposite side.* Once you have sewn the last LED, knot and cut the thread. Repeat, looping the thread several times around the *inner* battery slit on the right side of the felt. This will be your positive battery contact. Continue sewing a path to the *bottom* of the felt until you reach the edge of the rectangle cutout. Continue the path, securely sewing the positive leads to the felt. Once you have sewn the final LED, knot and cut the thread. *The conductive threads from the positive and negative LED leads should never touch.*

e. Wrap the felt around your wrist until both edges align. Slip the battery through the battery slit with the positive side of the battery touching the positive contact and the negative side of the battery touching the negative contact. Your LEDs should light up.

NOTE: Why don't I need a resistor?

Typically, when working with LEDs, you need to add a resistor to the circuit to prevent the LEDs from burning out. In this circuit, we are using conductive thread instead of wires to connect the 3 LEDs together in parallel and are using a 3V coin cell battery with a fairly minimal capacity. Unlike traditional wires, the conductive thread has a much higher level of resistance, and since the LEDs are rated for 3V, we don't need to add an additional resistor to keep the current at safe levels.

3. MAKE THE MAGNET CLASP

a. Next, place the felt wrong side up. Place the decorative fabric from Template D wrong side up, covering the rectangular cutout and LEDs. Using a sewing machine, topstitch along the rectangular cutout, sewing the decorative fabric securely in place.

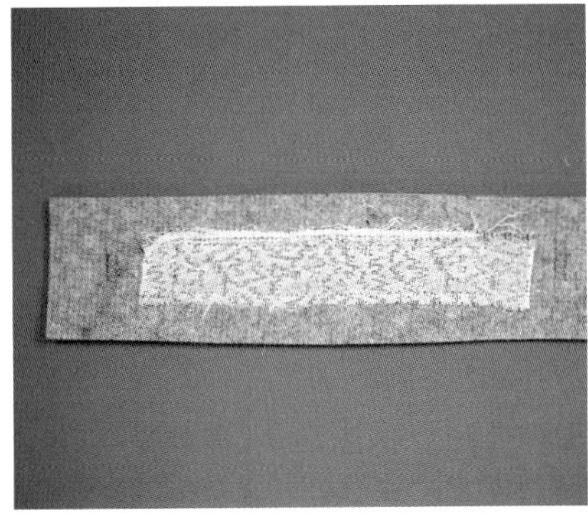

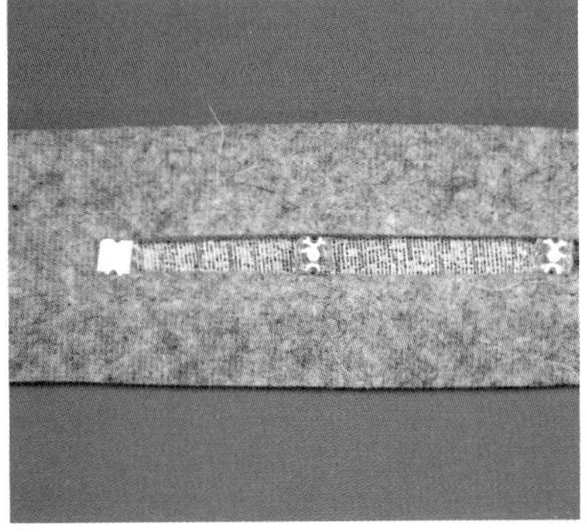

b. Using hot glue, glue a magnet at least ¼" from the edges to the top right corner of the felt. Repeat, gluing the second magnet to the bottom right corner.

c. Using scissors, cut the heat-shrink into four 1" pieces. Slip a piece of heat-shrink over the top and bottom leg of each cotter pin. Using a heat gun or hair dryer, shrink the tubing. Repeat for the second cotter pin. The heat-shrink will insulate the pin from the battery and prevent the circuit from shorting.

d. Align a cotter pin with the battery slit. The battery slit should be between the top and bottom legs of the pin. Using ordinary thread, sew the curved corner of the pin to the felt. Knot and cut the thread. Now sew only one of the pin legs to the felt. The cotter pin will help hold the coin cell battery in place. Repeat for the opposite side, gluing the remaining 2 magnets and sewing the cotter pin in place.

NOTE: Both sets of magnets on opposite sides should align and snap together (not repel each other). If the magnets are repelling one another, flip one of the magnets over.

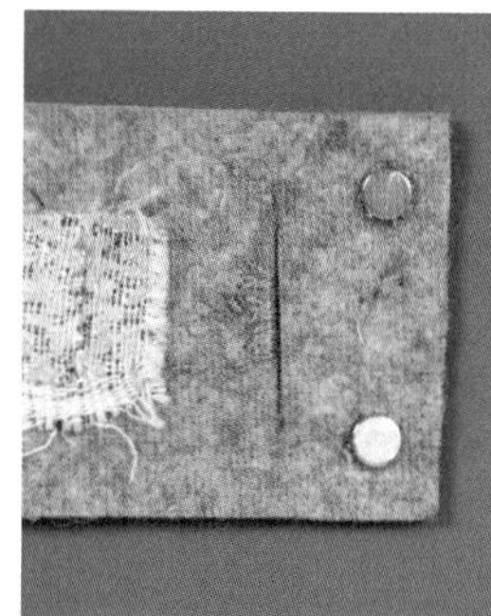

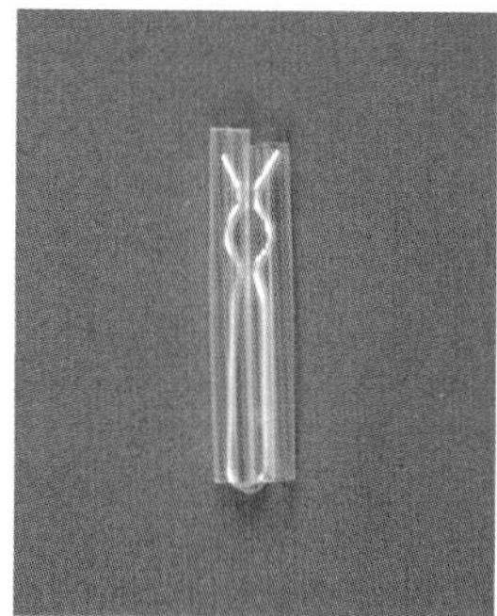

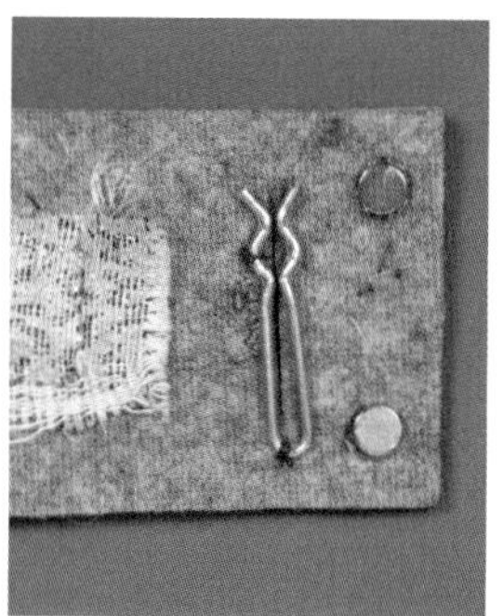

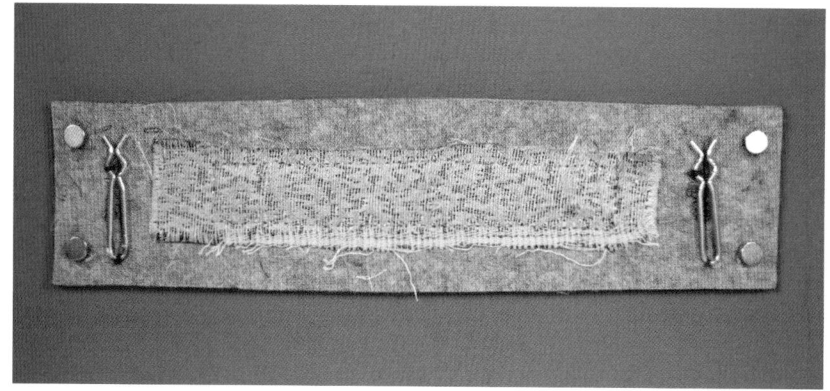

4. SEW THE BACK LINING AND BATTERY POCKET

a. Grab the back lining and battery pocket fabric. Mark the position of the right battery slit onto the back lining. Position the pocket piece at least ¼" to the left of the mark. Using a sewing machine, topstitch the pocket piece onto the back lining. The pocket piece will be used to store the coin cell battery when the bracelet is not being worn.

b. Place the industrial felt layer wrong side up. Place the back lining layer right side up on top of the industrial layer. Using a utility knife, slice the back lining layer in between the cotter pin so the coin cell battery can slip through.

NOTE: Depending on the fabric you are using for the lining, you may want to stitch around the slit area to prevent it from fraying. An alternative is to use Fray Check.

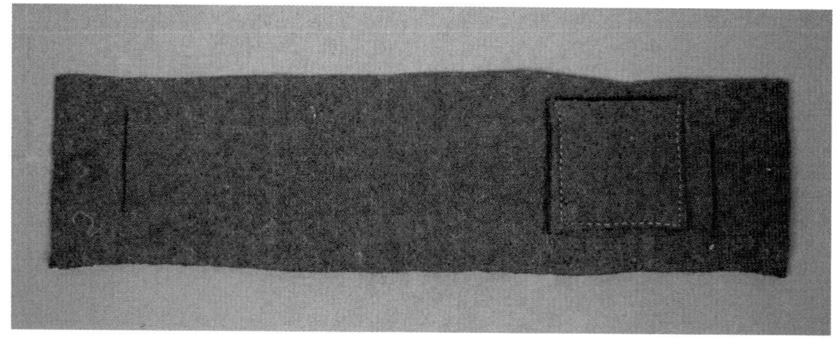

c. Using ordinary thread, hand-stitch the back lining piece to the industrial felt layer, sewing only along the surface of the felt and not entirely through the felt.

d. Put the cuff on and slip the battery through the battery slit with the positive side of thc battery touching the positive contact and the negative side of the battery touching the negative contact. You are now perfectly accessorized to be the light of any party!

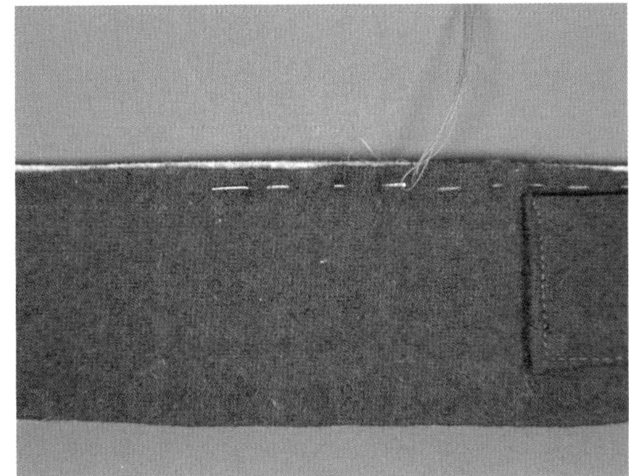

At the end of a long evening, don't forgot to store the battery in the inner pocket. The pocket is a handy feature that will help prevent you from misplacing the battery.

FINISH

ROCK STAR HEADPHONES

Keep the chill away by making gear that will keep your ears warm and protected from the harsh winter wind. The Rock Star Headphones are the perfect gear to keep you rockin' — and working out. Dressed with 2 small LEDs for a little added bling and safety, now you can hit the slopes or go for a brisk morning jog in style.

HOW IT WORKS »

The Rock Star Headphones require you to hack into an existing set of headphones and incorporate it into a new design. The Rock Star will also have two square LEDs sewn onto the exterior of the earwarmers. The LEDs have both an aesthetic function and a practical, safety function: making the wearer visible for jogging or biking at night. The circuit incorporates a soft switch made from conductive hook and loop, so the LEDs turn on only when the headphones are worn.

COMPLEXITY: BASIC

WHAT YOU WILL LEARN

- Work with LEDs
- Hack gear and work into projects
- Make a soft switch

RELATED TUTORIALS

- Working with LEDs (19-23)
- Sewing Soft Circuits (57-67)
- Measuring the Voltage of a Circuit (71)

MATERIALS

- **½yd fleece or sherpa suede fabric** for top fabric
- **½yd neoprene** for lining
- **2" conductive hook and loop**
- **Conductive thread**
- **Metal snaps (2)**
- **Thread**
- **Tape**
- **Embroidery thread**

TOOLS

- Scissors
- Sewing needle
- Needlenose pliers
- Pencil or marking pen
- Sewing machine
- Soldering iron and solder

ELECTRONICS

- **LEDs (2)** Piranha high-flux LEDs are recommended.
- **Headphones** Inexpensive ones work just fine.
- **Cellphone battery with leads** Goldmine-elec.com offers low prices.

TEMPLATES

- Template A
  Exterior suede and inner neoprene lining (1 each)
- Template B
  Pocket (1)

TEMPLATES

Templates available online at fashioningtechnology.com/headphones.

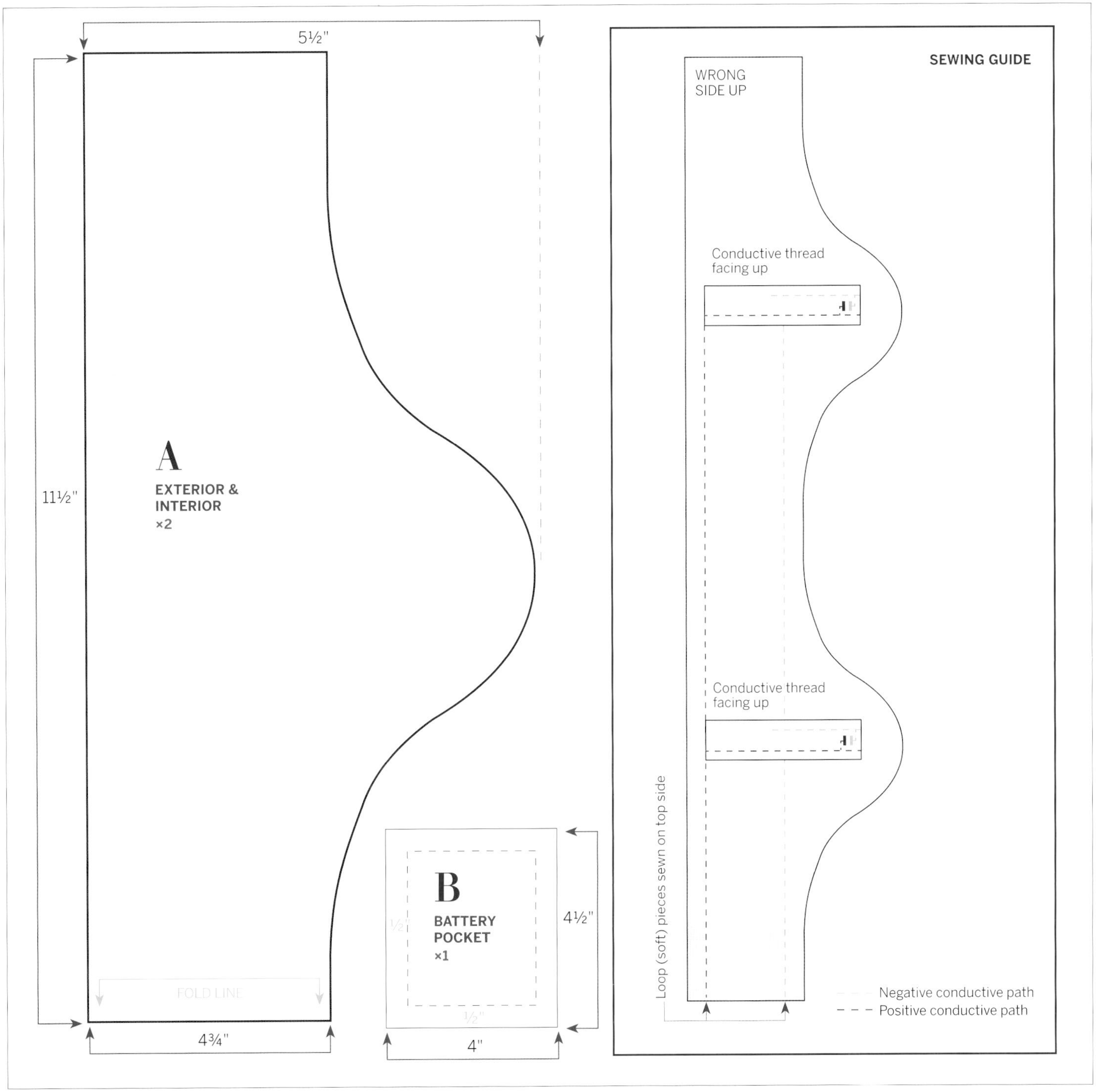

5½"

11½"

A

EXTERIOR & INTERIOR
×2

FOLD LINE

4¾"

B

BATTERY POCKET
×1

½"

½"

4½"

4"

SEWING GUIDE

WRONG SIDE UP

Conductive thread facing up

Conductive thread facing up

Loop (soft) pieces sewn on top side

Negative conductive path

Positive conductive path

START »

1. CUT THE TEMPLATE

Using a marking pen, trace Template A onto the interior of the suede. Repeat for the inner neoprene lining. Cut out the template. Then trace Template B onto the neoprene, and cut the template out. Set aside. This will be the battery pocket piece.

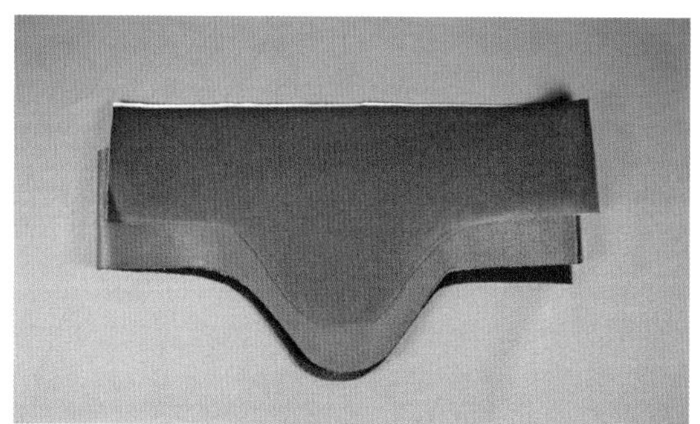

2. SEW THE CIRCUIT PATH

a. Trace the positive and negative conductive paths (from page 89) onto the interior of the suede. Place a bobbin of conductive thread in the sewing machine, and a spool of regular thread the same color as the suede in the spool pin.

b. Machine-stitch both conductive paths along the traces, making sure that the conductive thread is at the bottom of the fabric. Leave about 5" of loose thread at the beginning of the path. Knot and cut the extra thread at the end of the path. The upper conductive path will be the positive path. The lower conductive path will be the negative path.

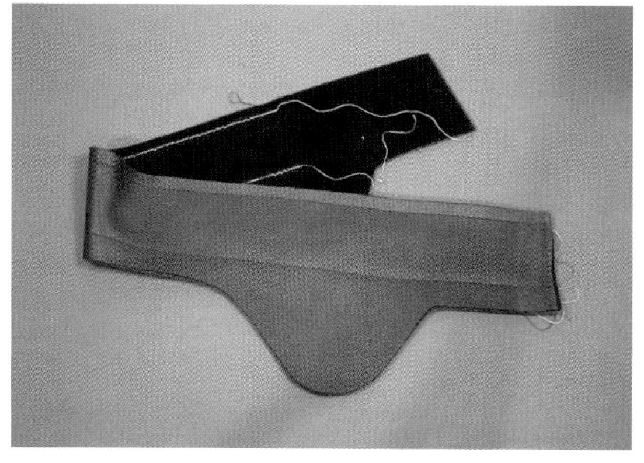

c. Replacing the conductive thread bobbin with ordinary thread, machine-stitch the loop pieces (soft side) of the hook and loop horizontally to the top and bottom right corners of the suede. The loop pieces should be sewn on the top of the suede.

d. Thread a sewing needle with the loose conductive thread from the negative path. Sew through the loop piece several times, making a connection between the negative path and the bottom loop. Knot and cut the conductive thread. Repeat for the upper loop piece, sewing the positive path to the upper loop.

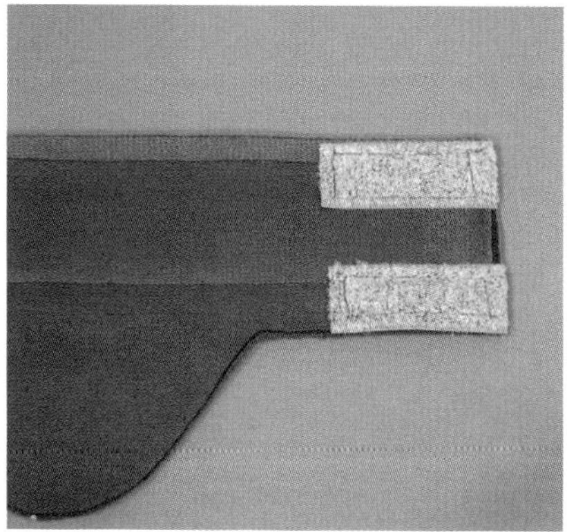

e. Cut two 1"×4" pieces of neoprene. Place the bobbin of conductive thread back into the machine.

Machine-stitch a 4" conductive path lengthwise about ¼" from the edge. Machine-stitch a 2" conductive path about ¼" from the opposite edge. Repeat for the second piece of neoprene.

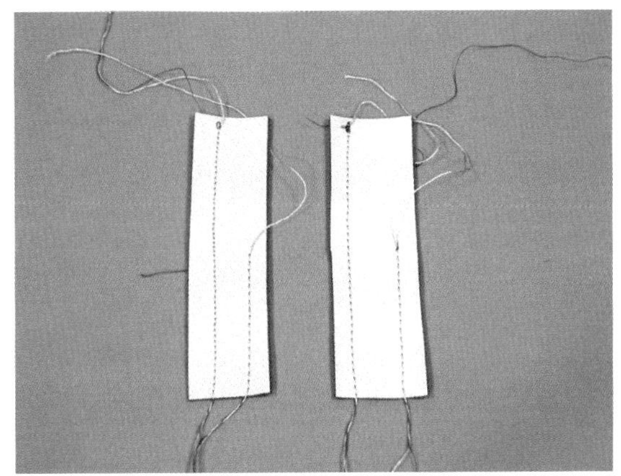

f. Pierce the leads of an LED with the positive leads facing up through the front of the suede at your desired location. Place the suede piece wrong side up. Grab one of the neoprene pieces. Pierce the leads of the LED through the neoprene in between the 2 conductive paths. The best way to determine which are the positive and negative leads is to refer to the datasheet or wire them temporarily using alligator clips (see page 29). The conductive path on the neoprene should be facing up. Using needlenose pliers, bend the leads flush to the fabric. Repeat for the opposite side.

g. Using conductive thread, hand-sew the positive lead of the LED to the long conductive path on the neoprene. Using a separate piece of conductive thread, sew the negative lead of the LED to the short conductive path on the neoprene. *Make sure that the conductive threads from the positive and negative LED leads never touch.*

h. Using conductive thread, hand-sew the long conductive path on the neoprene to the upper positive conductive path on the suede. Using a separate piece of conductive thread, sew the short conductive path on the neoprene to the lower negative conductive path on the suede. Repeat for the second LED. The positive lead of the LED should now be connected via the neoprene to the upper positive conductive path terminating at the positive loop piece and vice versa, with the negative lead connected via the neoprene to the lower negative conductive path terminating at the negative loop piece.

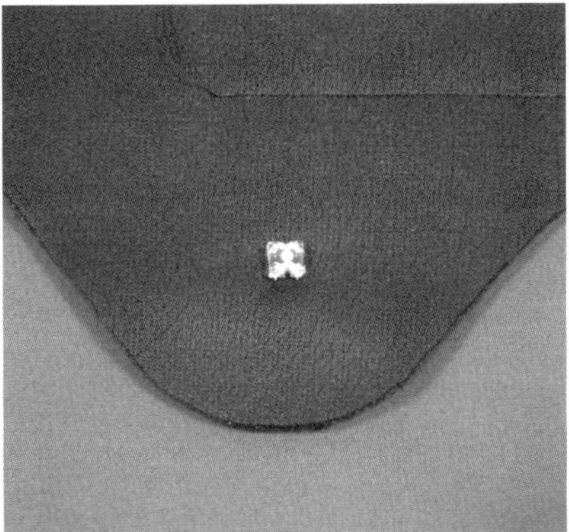

NOTE: In this example, we have used high-flux LEDs. High-flux LEDs have 4 legs — 2 positive (anodes) and 2 negative (cathodes). You can substitute the high-flux LEDs with typical LEDs. If you are using typical LEDs, curl the negative LED lead (the shorter one) into a loop. Mark it with a black marker to help you distinguish the negative lead from the positive. Then curl the positive lead.

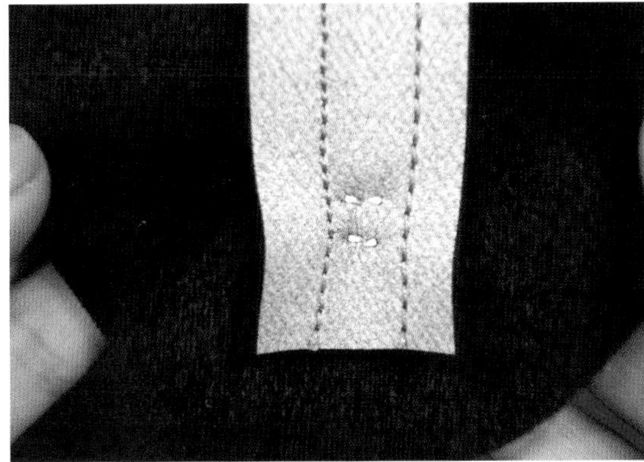

3. SEW THE BATTERY PACK

a. Take the cellphone battery, and solder a female snap to the positive lead and a male snap to the negative lead.

NOTE: **Some cellphone batteries have a third signal lead, typically coated in blue. If you can't distinguish the positive and negative leads (typically positive is red and negative is black), use a multimeter. Refer to page 71 for details on how to use multimeters.**

b. Place the neoprene right side up with the ear flap facing down. Using ordinary thread in the bobbin, machine-stitch the hook pieces (prickly side) of the hook and loop horizontally to the top and bottom right corner of the neoprene. Using conductive thread in the bobbin, sew a path about 1" to the left and 1" down (making an "L" shape) from the upper hook piece. Using conductive thread, end the path by sewing on a male snap.

c. Using conductive thread in the bobbin, sew another path about 1" to the left and 1" up (making an "L" shape) from the lower hook piece. Again using conductive thread, end the path by sewing on a female snap.

d. Fold the battery pocket piece along its hemlines. Machine-stitch the pocket piece next to the hook pieces, covering the conductive paths and snaps.

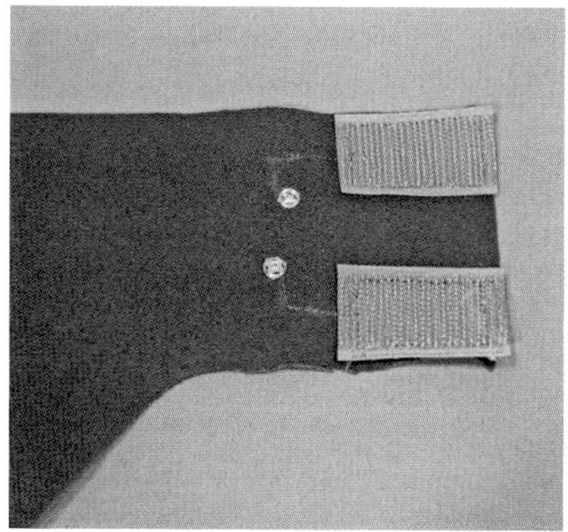

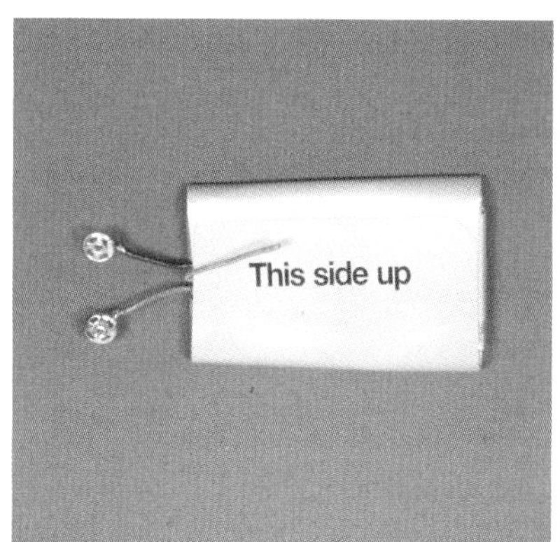

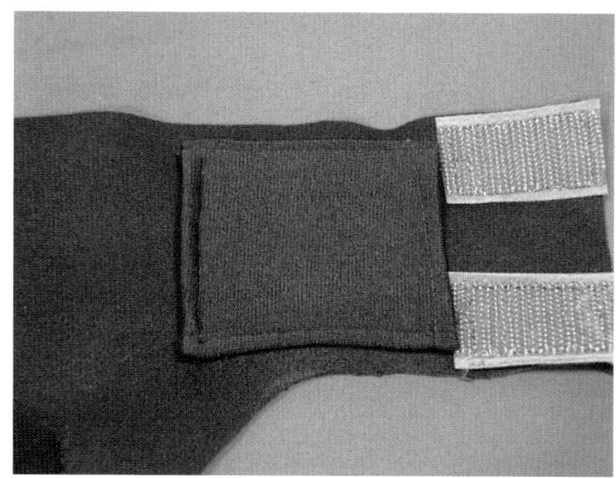

NOTE: The path must be connected to the hook piece. Begin the path inside the hook piece.

4. EMBROIDER THE HEADPHONES

Personalize your headphones by embroidering custom designs. For those not particularly skilled in the craft of embroidery, you can't go wrong with a simple five-point star. Using a tracing pen, trace a five-point star on top of the suede. Using embroidery thread, embroider the star. Repeat for the opposite side.

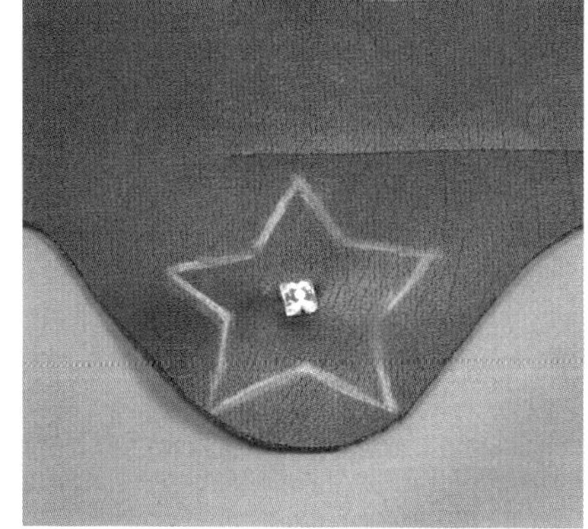

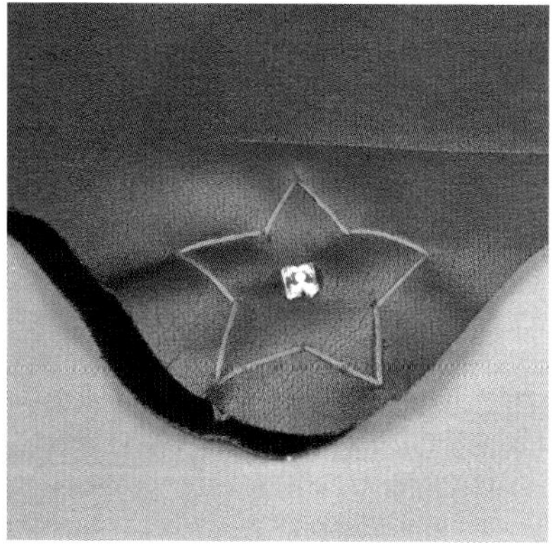

5. SEW THE HEADPHONES

a. Using needlenose pliers, carefully deconstruct the headphones until you have the 2 speakers loose from their housing.

b. Take the earpiece covers and place the neoprene piece right side up. Using ordinary thread, hand-stitch one of the earpiece covers to the center of the curved ear flap. Repeat for the other side.

c. Turn the neoprene over, placing it wrong side up. Using a utility knife, slice down the center of the back of the earpiece cover, being careful not to slice through the earpiece cover itself. Repeat for the second earpiece cover.

d. Slip the headphone earpieces into each slot. Adjust the wires so that the audio plug is located near the hook and loop, and extending from the bottom of the neoprene. You can temporarily hold the wires in place with tape. Once you have positioned the wires appropriately, use ordinary thread to hand-stitch sections of the wire in place.

e. Pin the neoprene lining to the upper suede piece. Machine-stitch the 2 pieces together, making sure to sew over (not through) the wires extending from the audio plug.

f. Snap the battery into place, grab your MP3 player, and begin rockin out to your favorite tunes — nice and toasty!

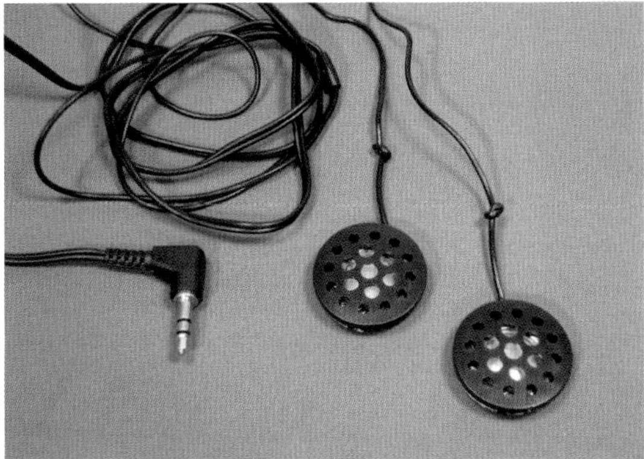

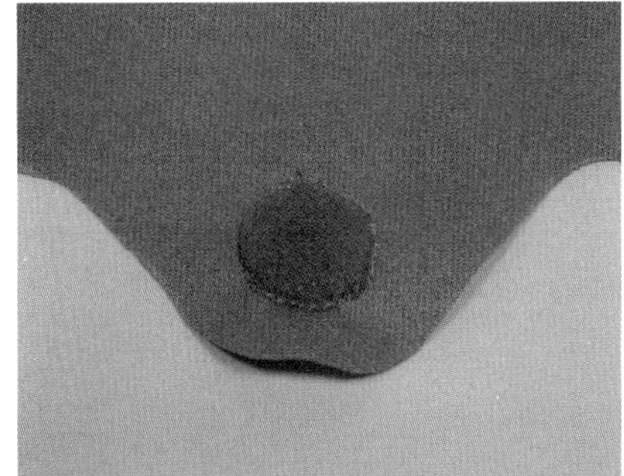

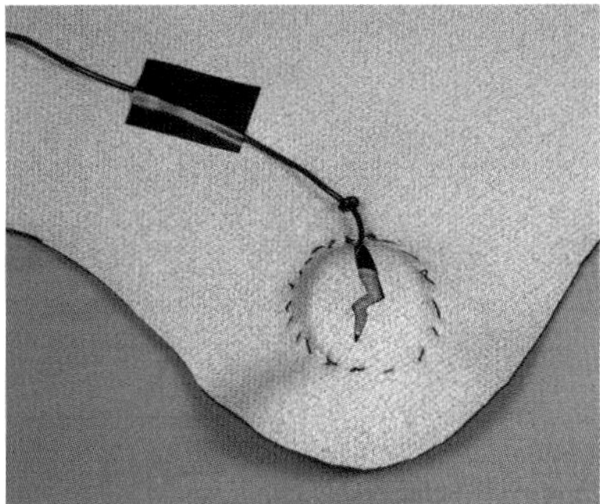

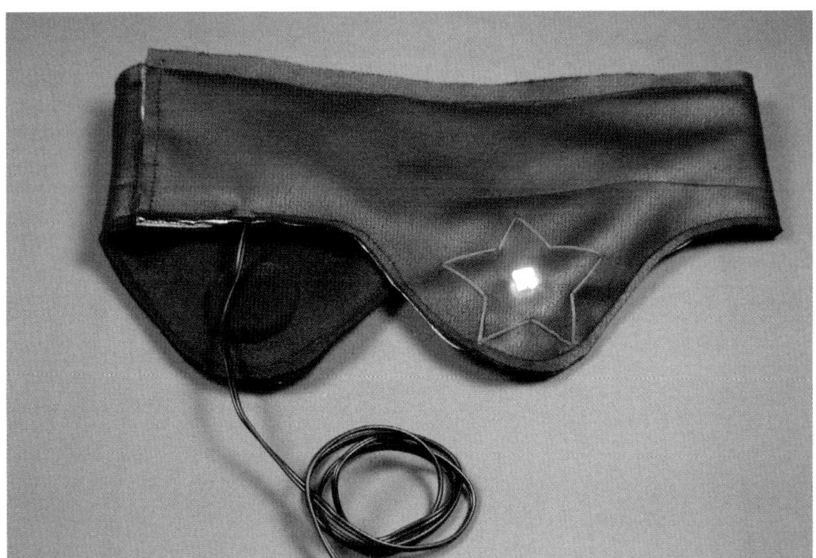

FINISH 

SPACE INVADERS TOTE

Accessorized with a handmade Space Invaders Tote for an evening shindig, you will never miss a call again. Whether you're out in a crowded, loud bar or trying to be discrete at a friend's poetry reading, the Space Invaders Tote provides an ambient light signal letting you know that someone is trying to reach you.

HOW IT WORKS »

The Space Invaders are sensitive to the electromagnetic waves of mobile phones. When in close proximity to a mobile phone, the Space Invader is energized by an incoming call and its LED eyes begin to twinkle.

COMPLEXITY: INTERMEDIATE

WHAT YOU WILL LEARN

» Solder and desolder
» Sew a soft circuit
» Work with LEDs

RELATED TUTORIALS

» Working with LEDs (19-23)
» Sewing Soft Circuits (57-67)
» Art of Soldering (45-49)
» Desoldering (49)

MATERIALS

» **Metallic snaps (4)**
» **1yd sturdy fabric** for exterior of bag
» **1yd lining fabric**
» **½yd metallic nylon ripstop fabric** for Space Invader
» **½yd ironing board material** for top layer of Space Invader
» **1⅓yd nylon webbing** for shoulder strap
» **Interfacing**
» **Metal strap carriers (2)**
» **Metallic thread**
» **Nonmetallic thread**
» **Fabric fusing material**
» **Conductive thread**
» **Electrical tape**
» **Ruler**

TOOLS

» **Sewing machine**
» **Soldering iron**
» **Third hand** (recommended)
» **Tracing pen**
» **Sewing pins**
» **Needle**
» **Scissors**
» **Needlenose pliers**
» **Utility knife**
» **Iron and ironing board**
» **Wire strippers**
» **Rosin-core solder**
» **Desoldering braid**

ELECTRONICS

» **Cellphone flasher** part #PF1 001 mutr.co.uk
» **26 AWG stranded wire, 12"**
» **LEDs (2)** Piranha high-flux LEDs are recommended.

TEMPLATES

» **Template A**
Space Invader bottom (2) and interfacing (2)
» **Template B**
Space Invader top (2)
» **Template C**
Eyes (4)
» **Template D**
Bag exterior (1) and bag lining (1)
» **Template E**
Inner pocket (1)
» **Template F**
Outer pocket (1)
» **Template G**
Strap handle (2)

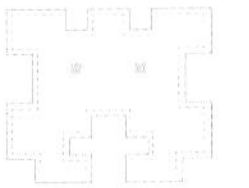

TEMPLATES

Templates available online at fashioningtechnology.com/spaceinvaders.

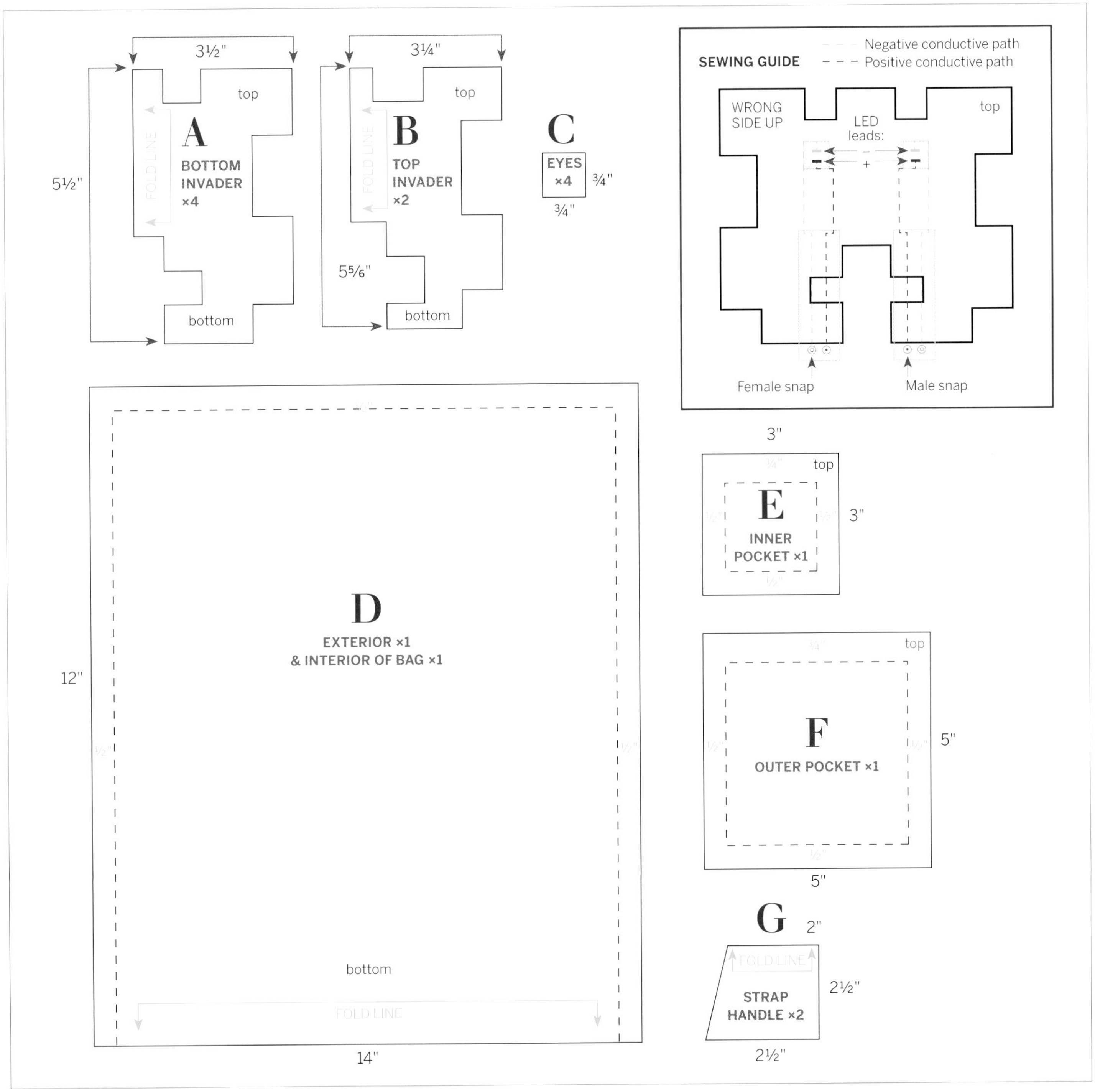

START »

1. WIRE THE PHONE FLASHER UNIT

The phone flasher comes with 2 LEDs presoldered on the circuit board. You will be removing the LEDs and replacing them with wires soldered with a pair of snaps at each end. A third hand tool is recommended.

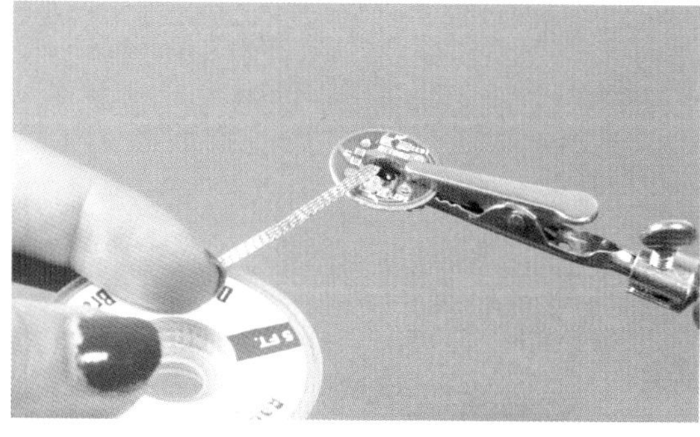

NOTE: Before removing LEDs from phone flasher unit, test the unit first (by checking your voicemail, for example) to ensure that it works with your particular cellphone and/or that the batteries are not discharged. If the LEDs do not blink, try replacing the batteries.

a. Remove the batteries from the phone flasher board.

b. Using a desoldering braid, place the braid over a solder joint of an LED. Using the iron tip, press the tip over the braid and joint. Once the solder is molten, remove the braid. (For a detailed tutorial on how to desolder, refer to page 49.) Repeat for the second joint, completely removing the LED from the circuit board. Repeat for the second LED. Both LEDs should now be removed.

c. Cut and strip the ends of four 1½" pieces of stranded wire. Twist the ends tightly so that they don't fray. Next, tape the phone flasher board onto a hard surface (or use your third hand tool), making sure to leave the solder points open.

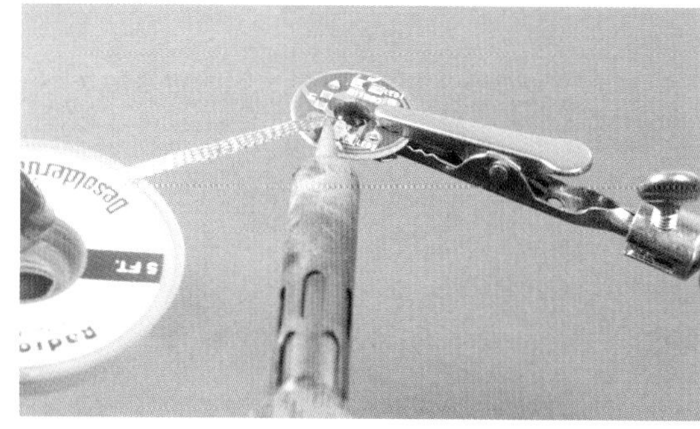

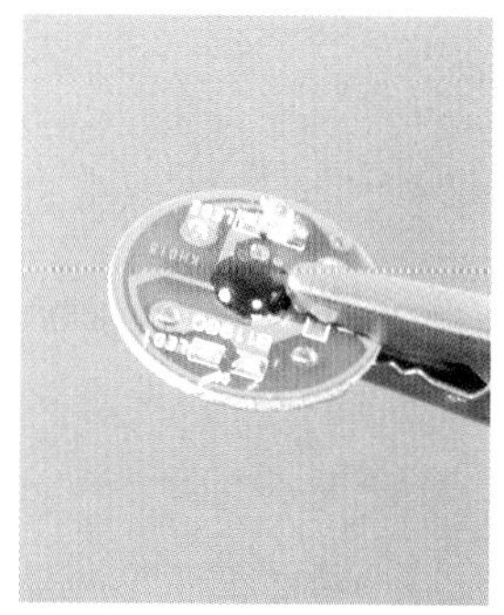

d. Solder one end of the stranded wire onto one of the solder points. Repeat for the other 3. The wires soldered near the markings "LED1" and "LED2" on the board will be the negative wires, and the other 2 will be the positive. Using a black marker, mark the negative wires.

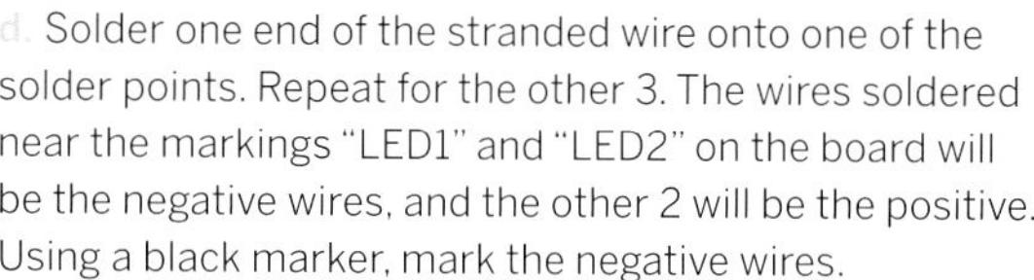

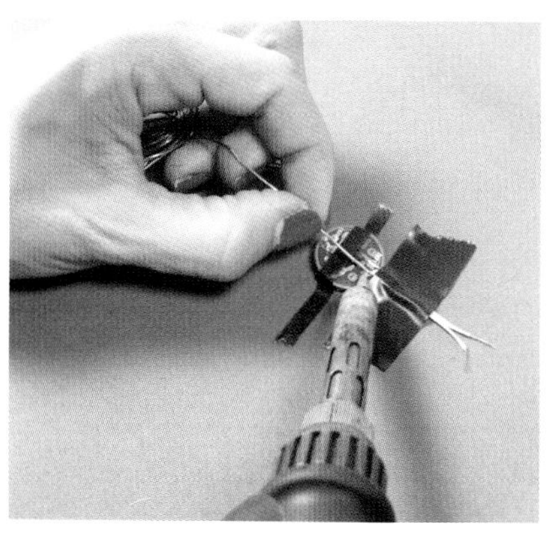

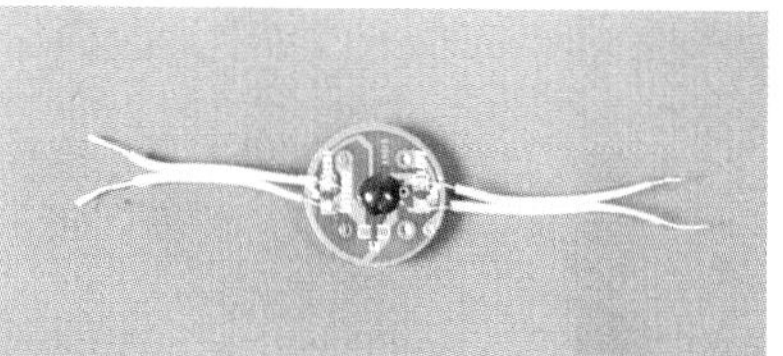

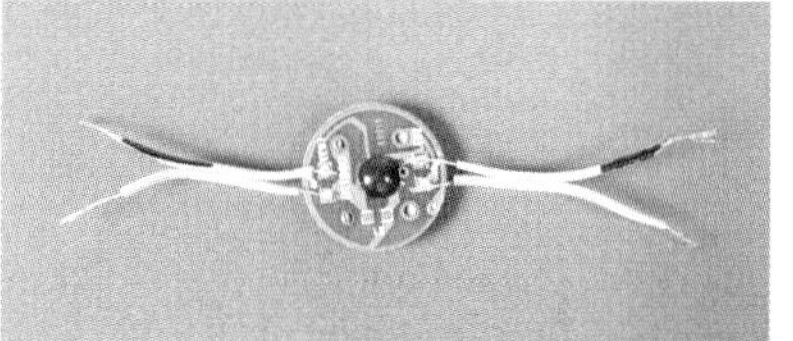

e. Grab the female ends of the snaps. Loop and twist one of the positive wires through a female snap. Add a touch of solder to secure the connection. Repeat for the other positive wire and remaining female snap. Grab the male ends of the snaps. Repeat the process, soldering the negative wires to the male snaps.

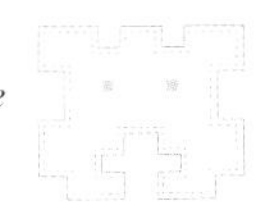

2. CUT THE SPACE INVADER TEMPLATE

Use the Space Invader template or create one of your own. You will be making 2 Space Invaders to sew onto the upper portion of your tote, one with a pair of LEDs for its eyes.

a. Cut the metallic nylon ripstop fabric into four 6"×8" pieces. Fold the 6"×8" fabric in half, lengthwise, wrong side out, aligning edges. Pin the side of Template A marked "fold line" to the folded edge of the fabric. Template A will be the bottom layer of the Space Invader — the backing.

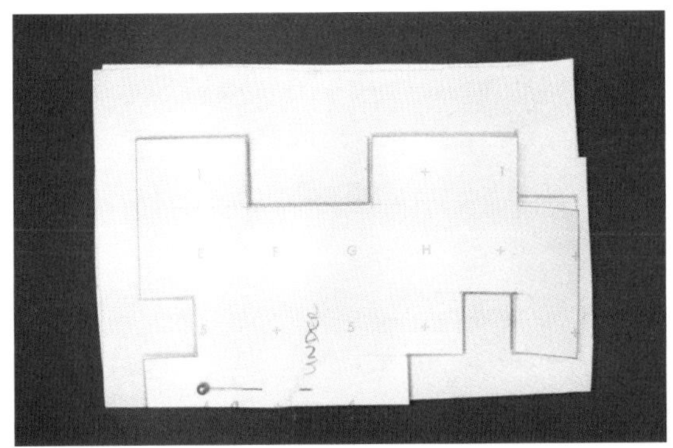

b. Using a tracing pen, trace the pattern onto the fabric. Cut the fabric along your traced lines. Repeat 3 more times, cutting a total of 4 pieces.

c. Repeat Steps 2a and 2b for Template B using the ironing board fabric, cutting a total of 2 pieces. Template B will be the top layer of the Space Invader.

d. Pin Template C onto the metallic nylon ripstop fabric and trace the template. Cut a total of 4 pieces. Template C will be the eyes.

e. Using Template A, trace and cut 2 pieces of interfacing slighter smaller than the template. The interfacing will be sandwiched between the top and bottom fabric pieces to give the Space Invader structure.

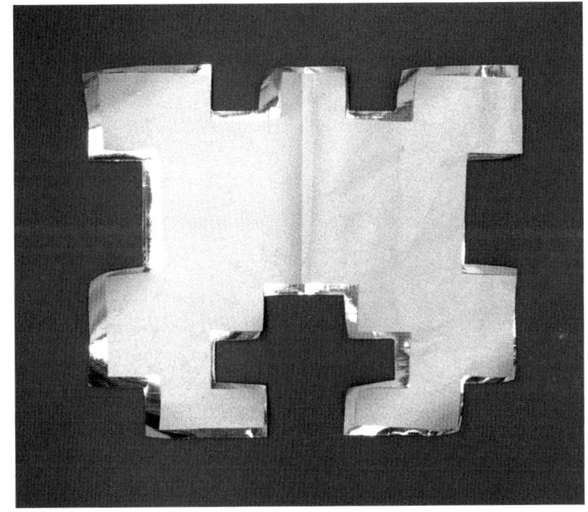

3. SEW THE SPACE INVADERS

a. Topstitch the upper fabric layer of the Space Invader to the bottom layer. Repeat for the second Space Invader.

b. Topstitch the eyes on top in any desired location, preferably at least 1½" apart. This will be the front of the Space Invader. Repeat for the second Space Invader.

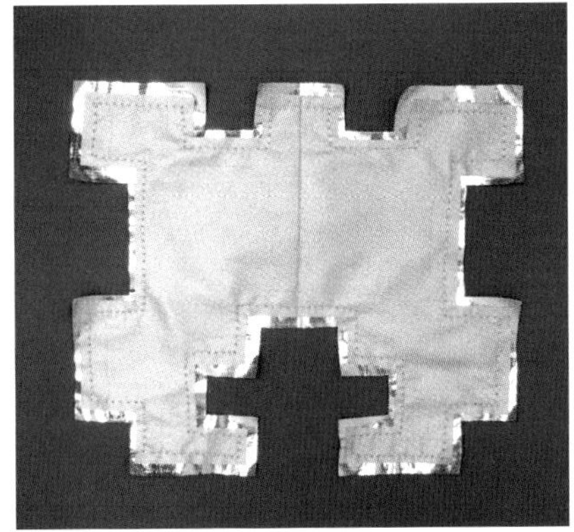

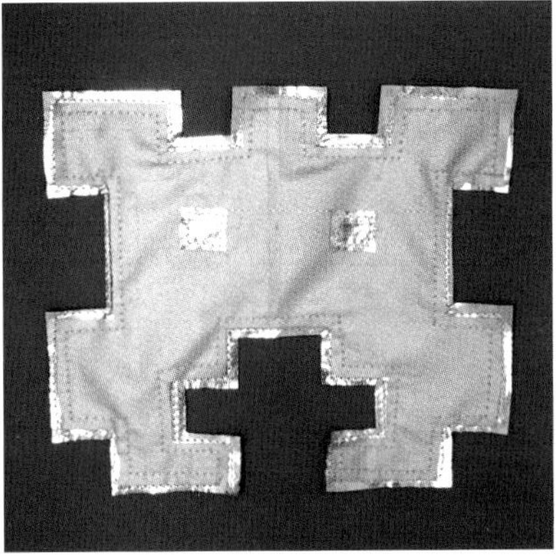

c. Take one of the interfacing pieces. Sandwich the interface layer between the top and bottom layers of the Space Invader. You can use a layer of fusing material to adhere the pieces together before sewing. Topstitch the 3 pieces together.

d. Mark the location of the eyes on the remaining interfacing piece. Using conductive thread, hand-stitch a line from the top outer section of an eye down to the bottom of the interfacing. This will be the negative line. Then hand-stitch a line from the bottom inner section of the eye down to the bottom of the interfacing. This will be the positive line. Repeat for the second eye. See sewing guide on page 97.

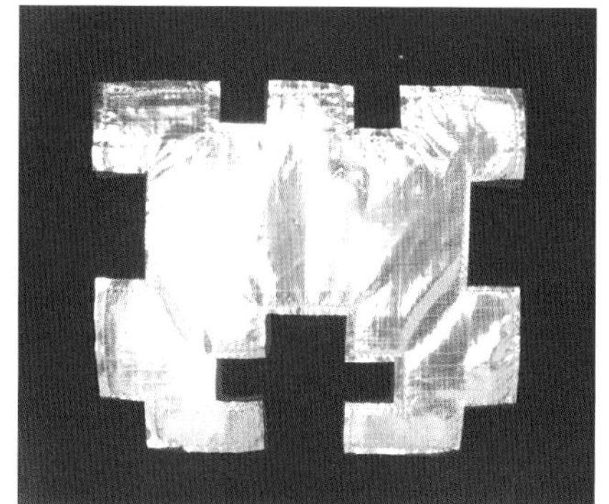

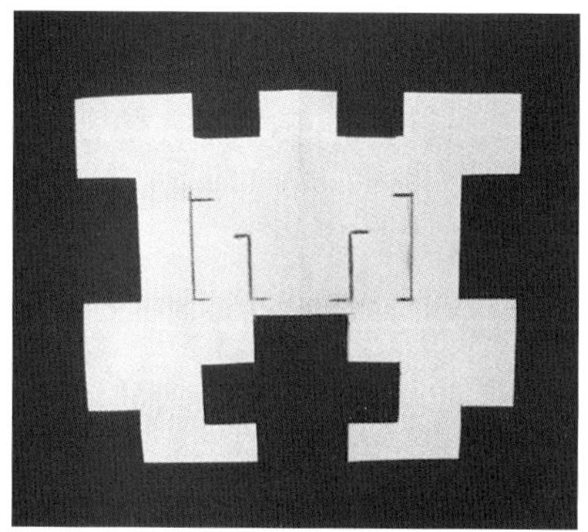

e. Cut two 1"×2" pieces of fabric. Starting ¼" from the left side, using conductive thread, machine-sew a line from the top of the fabric, stopping ¼" from the bottom. Leave 5" of loose thread from the top and bottom. Then sew a parallel line ¼" from the right side. Repeat for the second piece.

f. Thread a sewing needle with the loose thread from the bottom left of the fabric. Hand-sew a female snap, securing it to the fabric. Knot and cut off any extra thread. The female snap will be the negative snap. Repeat, sewing a male snap to the bottom right of the fabric. The male snap will be the positive snap. Repeat for the second fabric piece, this time swapping the placement of the snaps.

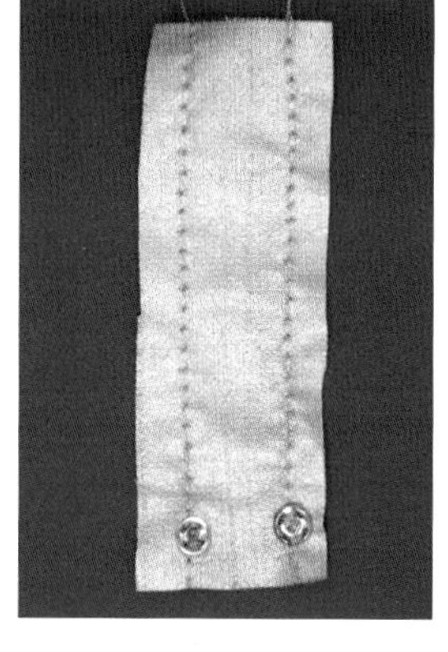

NOTE: For the second fabric piece, swap the placement of the male and female snaps so that the female snap is at the bottom left and the male snap is at the bottom right.

g. Grab the interfacing layer. Place a fabric piece face down, aligning the stitch lines from the positive (male) and negative (female) snaps to the positive (from bottom of eye) and negative lines (from top of eye) sewn on the interfacing. Refer to the sewing guide on page 97. Thread a sewing needle with the positive loose thread from the fabric piece. Securely sew the fabric piece onto the interfacing, overlapping the conducting lines from both pieces, creating a conductive path from the interfacing to the snap.

h. Thread a needle with the negative loose thread from the fabric piece. Securely sew the fabric piece onto the interfacing, making sure both negative conductive lines overlap. Repeat for the second fabric piece, aligning the stitch lines from the positive and negative lines accordingly. Place the combined top fabric layer face down. Place the interfacing layer directly on top of the fabric layer with the snaps facing down. Pin.

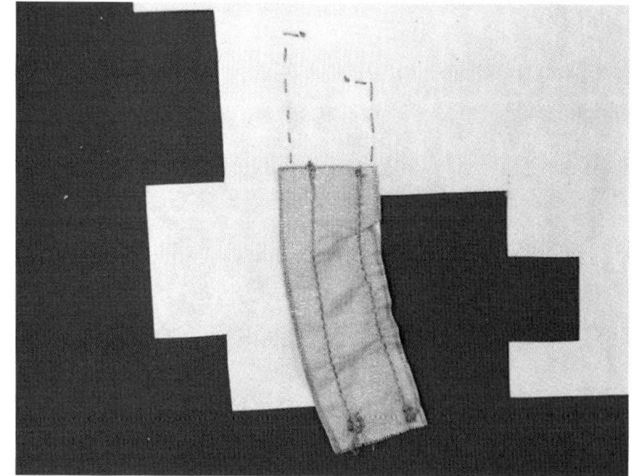

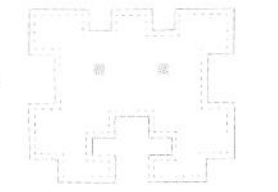

4. ADD THE LEDS

a. Turn the combined top fabric layer and interfacing layer right side up. Position an LED in the center of the eye square with the negative leads facing up. (If you are not sure which is the negative or positive lead of the LED, refer to page 74 to learn how to distinguish between the 2 leads.)

Pierce the leads of an LED through both pieces, having the leads exit through the top of the interfacing. Using a pair of needlenose pliers, gently bend the leads flush to the fabric.

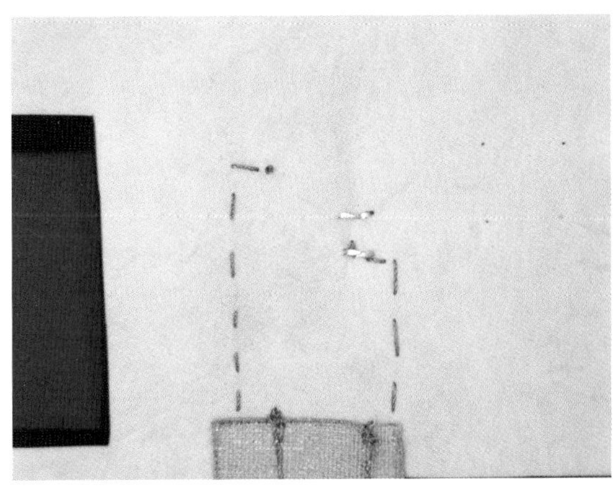

b. Using conductive thread, hand-sew the bottom positive leads of the LED to the positive stitched line on the interfacing. Be careful not to sew through the top layer. Repeat for the top negative leads, sewing them to the negative stitched line on the interfacing.

c. Repeat Steps 4a and 4b for the second LED. The positive leads of the LEDs should now be connected to the positive stitched line of the interface concluding at the positive male snap, and vice versa for the negative leads. All the positive conductive paths should now be connected together and all the negative paths to each other.

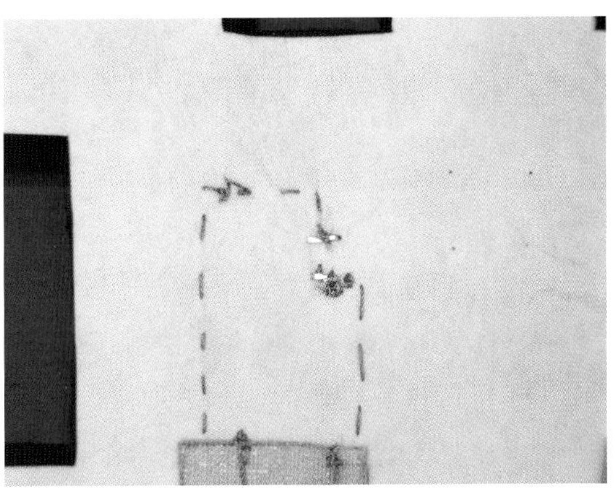

d. Align the bottom fabric layer on top of the Space Invader. Pin. Using scissors, cut 2" off the bottom fabric, creating an opening for the snap fabric strips. Topstitch the pieces together.

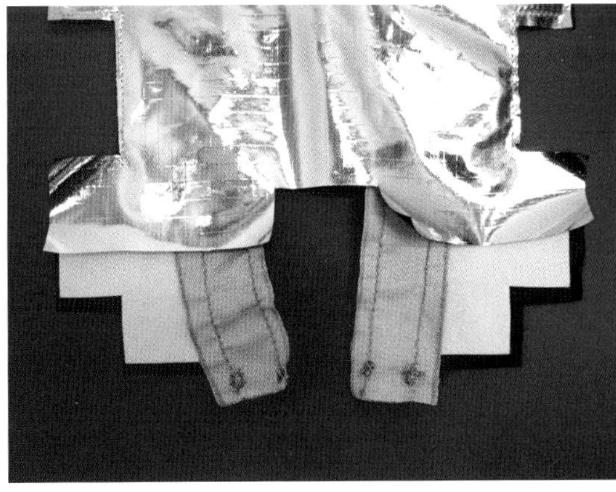

5. CUT THE BAG EXTERIOR AND LINING TEMPLATES

a. Fold the exterior bag fabric in half, lengthwise, wrong side out, aligning edges. Pin the side of Template D marked "fold line" to the folded edge of the fabric. Using a tracing pen, trace the template onto the fabric. Cut the fabric along traced lines, cutting notches.

b. Mark a ½" seam line along all sides except the fold line. Fold the fabric in half along the fold line, wrong side out, aligning edges and notch marks. Machine-stitch along both side seam line markings.

c. Repeat Steps 5a and 5b for the lining fabric, but do not machine-stitch yet.

6. ATTACH THE POCKET TO THE LINING

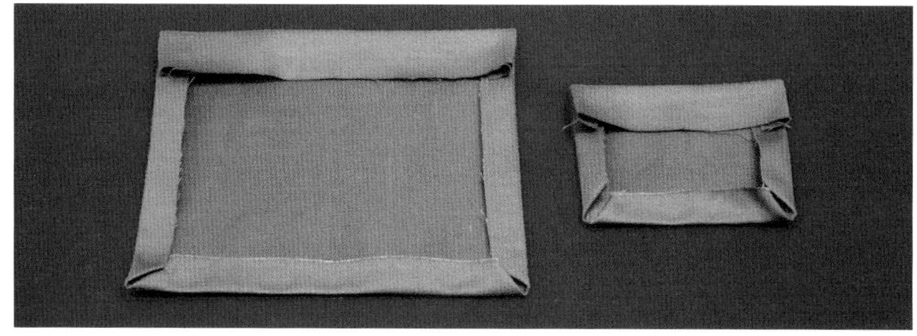

a. Pin Template E and F onto the fabric. Trace, including the seam lines. Cut the pocket pieces.

b. Place one of pocket pieces wrong side up on an ironing board. Fold the seam lines inward and press. Repeat for second pocket piece.

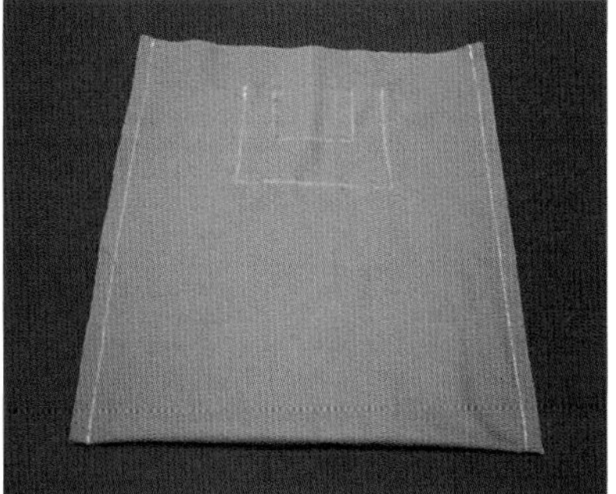

c. Machine-stitch along the ¾" seam for both pockets. The ¾" seam will be the top of the pockets.

d. Place the lining piece right side up. Center the small pocket piece, also right side up, 2" below the upper seam line. Topstitch the pocket piece into place. Repeat for the large pocket piece, aligning it 1¾" from the upper seam line. The large pocket piece will cover the small, inner pocket piece.

e. Fold the lining fabric in half along the fold line, wrong side out, aligning edges. Machine-stitch along both side seam line markings.

7. MAKE AND ATTACH THE STRAP HANDLES

a. Fold the strap handle fabric (should be similar to the exterior fabric) in half, wrong side out. Pin the side of Template G marked "fold line" to the folded edge of the fabric. Trace and cut the template along the marking lines. Repeat cutting for a total of 2 pieces.

b. Slip the strap handle piece through a metal strap carrier. Machine-stitch ¼" from ends. Repeat for the second strap handle.

c. On the exterior bag fabric, position the strap carriers at opposite ends and opposite sides of the bag. Align the bottom of the strap carrier piece to the top of the bag, with the angled side facing in, and pin to the outside of the bag. Machine-stitch the strap carriers, securing them in place.

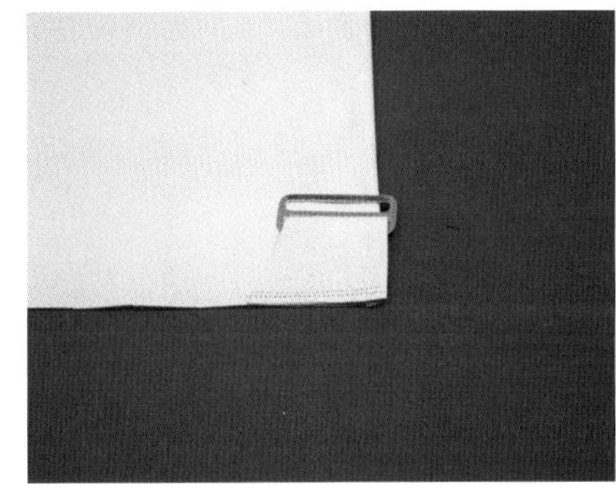

8. INSERT BAG LINING

a. Insert lining piece, right side out, into the exterior bag piece. Match the side seams. Then fold the ½" seam allowances of both the inner and outer pieces toward each other and pin. Press and topstitch around the bag about ¼" from the edge (the top seam, not the sides).

b. Make two 1½" horizontal buttonholes 2" apart from each other, locating them 1" from the top of the large inner pocket. These buttonholes will be used to slip the snap fabric pieces through to the interior of the bag.

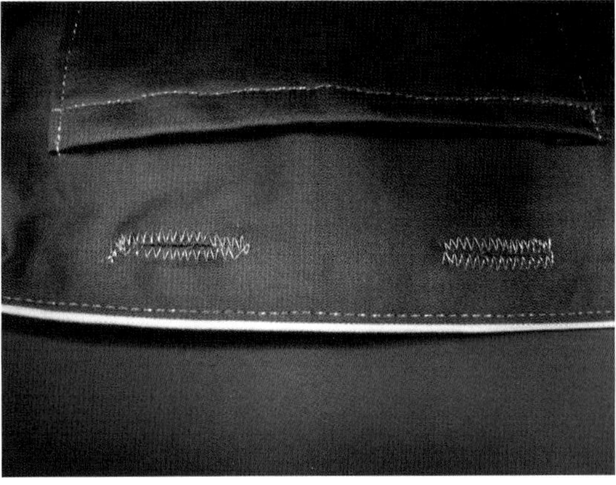

9. SEW THE SPACE INVADERS

a. Center the Space Invader *without the LEDs* on the bag, with 3" of it hanging off the top of the bag. The first Space Invader should be placed on the side of the bag *without the pockets*. Pin, then topstitch the Space Invader into place.

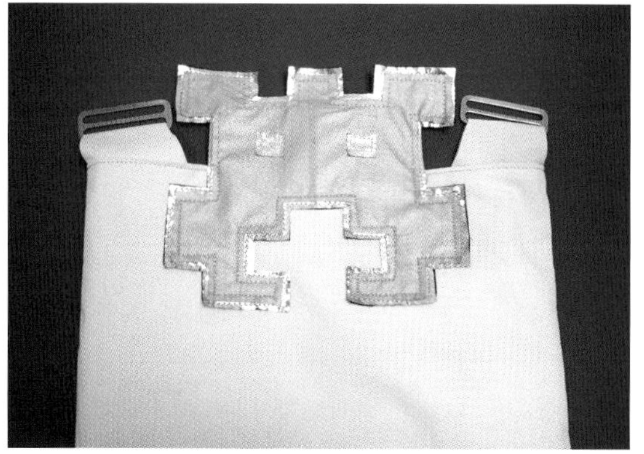

b. Center the second Space Invader *with* the LEDs on the opposite side of the bag, aligning it with the first Invader. Pin. Slip the snap fabric pieces through the buttonholes into the interior of the bag. Then topstitch the Space Invader, avoiding the interior pockets.

c. Snap the phone flasher into place and slip it into the small interior pocket.

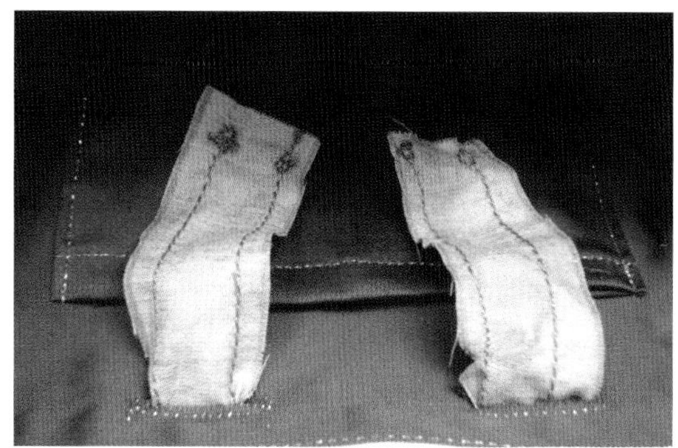

10. SEW ON THE SHOULDER STRAP

Cut the nylon webbing to your desired length. Slip the webbing through the strap handle, looping it over onto itself, and machine-stitch. Repeat, attaching the opposite end of the webbing to the second strap handle.

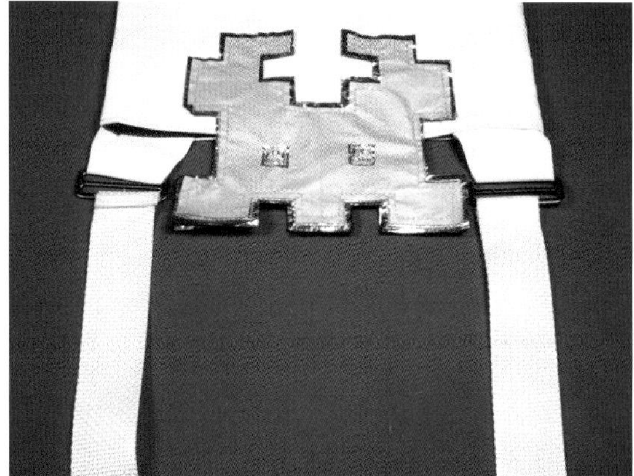

FINISH

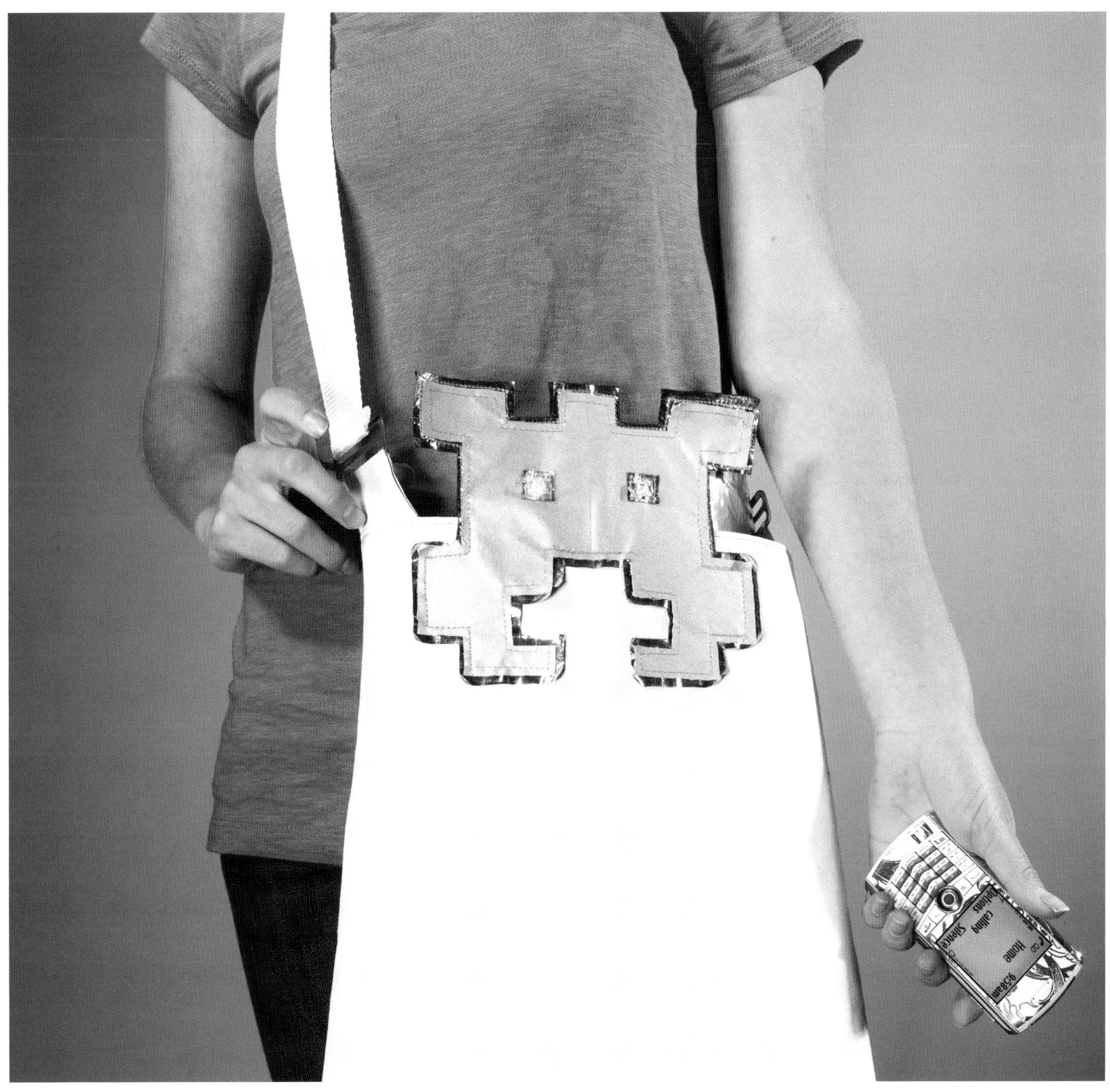

AERIAL *the* BIRDIE BROOCH

Perched on your lapel or messenger bag, Aerial the Birdie Brooch chirps and whistles, continuously changing her song. Her mood is affected by sunlight, tending to be more "chirpy" when it is bright and quieter in the shade. Aerial's circuit can be modified to create a custom song, giving each new avian companion you make a unique voice. Create your own personal choir by customizing both the aesthetic design and the pitch and frequency of each bird's song.

HOW IT WORKS »

Aerial's circuit is energized by a flexible solar panel, producing variable sounds. To create a unique sound, you should first build out the circuit on a breadboard. By experimenting with the circuit, you can change the pitch and frequency of her chirps by swapping different value capacitors (ranging from 1μF to 47μF) and resistors (ranging from 1K to 100KΩ). Once you have found a song you like, use those specific components to build your final circuit. If you can't decide between a few, build out several boards and create a wearable avian choir.

COMPLEXITY: ADVANCED

WHAT YOU WILL LEARN

- Etch a printed circuit board
- Work with ICs
- Use a breadboard
- Solder

RELATED TUTORIALS

- Art of Soldering (45-49)
- Building a Simple Circuit (27-33)
- Measuring the Voltage of a Circuit (71)

MATERIALS

- 2"×3" copper board
- PCB etchant solution
- Stranded wire, 8"
- Glossy inkjet photo paper or PCB transfer film
- 3/32" heat-shrink tubing, 2"
- Double-sided tape
- Brooch pin with adhesive backing

TOOLS

- Black permanent marker
- **Plastic trays (2)** Do not use metal.
- **Plastic or wooden tongs** with no metal
- Rubber gloves
- Drill and 1mm drill bit
- Clamps
- Steel wool (0 grade) or nylon abrasive pad
- Lint-free fabric
- Soldering iron and solder
- Scissors
- Wire cutters
- Wire strippers
- Iron
- Heat gun, hair dryer, or lighter
- **Toothbrush** (optional)

ELECTRONICS

- **IC Hex-Schmitt inverter**, generic, 74HC14, part #45364 from jameco.com
- **Capacitors (3)** ranging from 1μF to 47μF; ceramic disc or electrolytic
- **Resistors (3)** ranging from 1KΩ to 100KΩ, 1/6W–1/4W
- **Piezo speaker** Allelectronics.com has a good selection.
- **Flexible solar cell** minimum voltage > 2.5V
- **Breadboard** (optional)

PATTERN & TEMPLATE

Pattern, template, and circuit available at fashioningtechnology.com/aerial.

OWL PATTERN

If you would like to use the Owl for your design, download the Owl Pattern in addition to the circuit template.

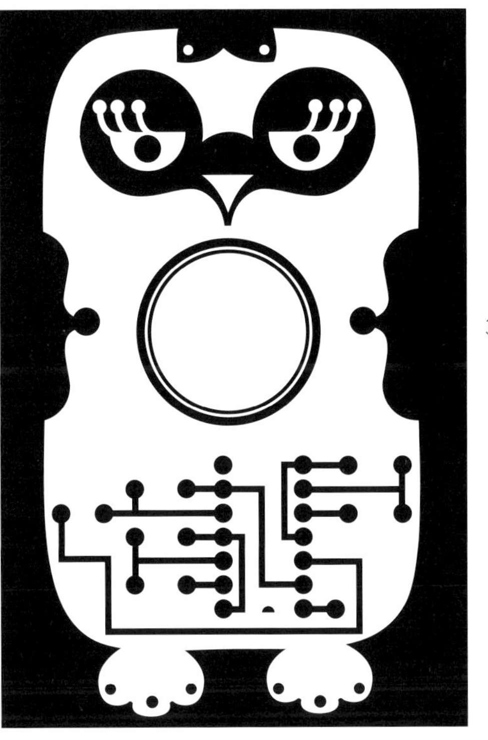

3"

2"

CIRCUIT TEMPLATE

If you would like to design your own pattern, you can download the circuit template only.

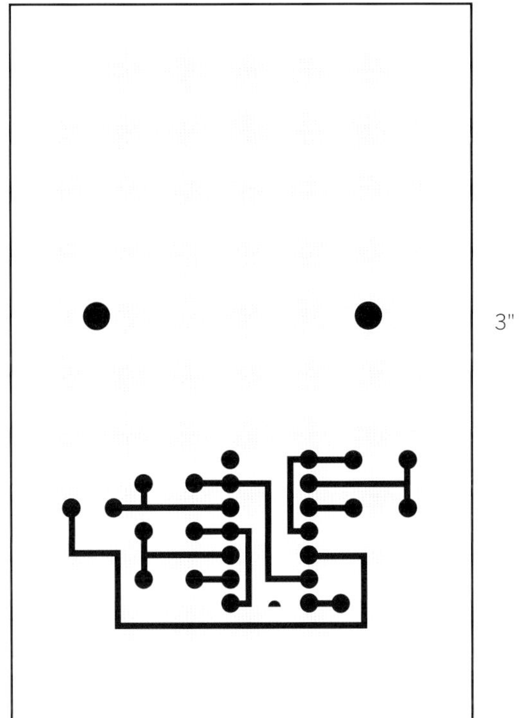

3"

2"

A special thanks to Ralf Schreiber for the design of this unique circuit.

START »

1. CREATE THE CIRCUIT BOARD PATTERN

You can create the Birdie Brooch by either downloading the pattern or by designing your own. Skip to Step 1b if you are not designing your own custom pattern.

a. Using the provided template, design your pattern around the circuit layout. Your pattern must be in black. You can create your pattern in one of 2 ways:

Option 1: If you know how to use any graphics software, you can bring the template into the desired software program and create your pattern digitally. Once your pattern is created, you *must* flip it horizontally (mirror) so it will transfer properly.

Option 2: Print out the template. Collage your pattern onto the template, making sure not to collage over the provided circuit layout. Photocopy your collage with the darkest settings. You must take into consideration that your pattern will transfer onto the copper plate flipped horizontally.

b. Print or photocopy the pattern onto glossy photo paper or PCB transfer film using the darkest laser printer or copier settings. Using a pair of scissors, cut out the pattern, leaving at least ¼" extra paper on one side.

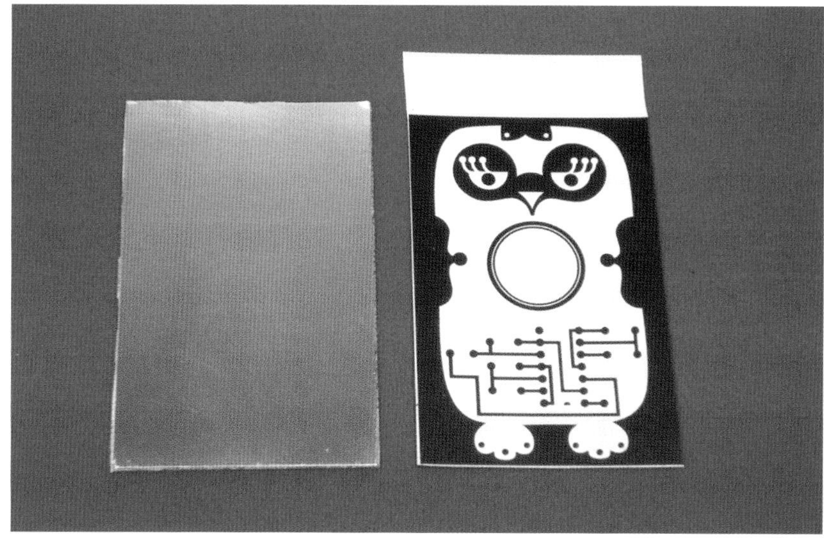

NOTE: The final pattern ***must be black***! If any areas are gray, they will not transfer onto the copper plate properly. The etchant will remove all the white areas of your pattern, and the black areas will remain.

Inkjet printers will ***not*** work. Use a laser printer or a photocopier to render your art.

2. TRANSFER THE IMAGE ONTO THE COPPER BOARD

a. Scrub the board lengthwise with a nylon abrasive pad or steel wool. This will create a rough surface for the toner to transfer. After scrubbing, wash the board with dish soap and water and dry it with a lint-free fabric.

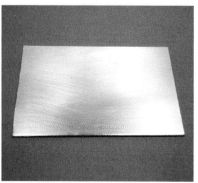

b. Lay the copper board (with the copper side facing up) on a heat-resistant surface such as a smooth piece of plywood. Align the printed pattern facing down along the edges of the copper board. You should have an extra ¼" of paper hanging off one side of the board.

c. Preheat the iron to the highest setting. Make sure to turn the steam off. Place the iron on top of the board and printed pattern, heating the entire board for about 30 seconds. For another minute or so, using the bottom edge of the iron, slowly move the iron over the entire pattern. You should begin to see the pattern showing through the back side of the paper.

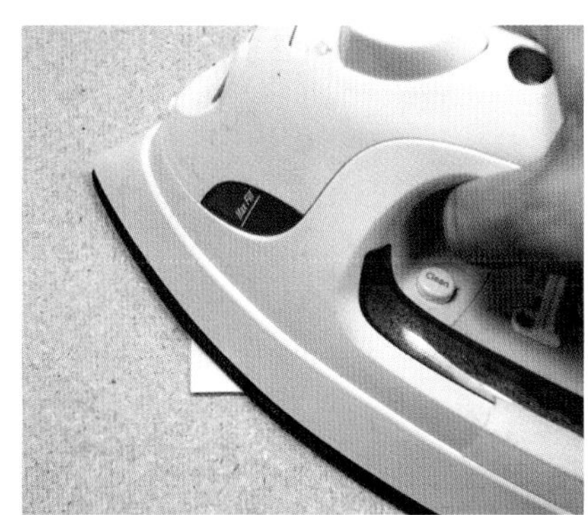

d. Fill a plastic tray with hot water (not boiling). Grab the circuit board with a lint-free cloth (the circuit board will be hot — use caution!) and drop it into the hot water for about 1–2 minutes.

e. Peel off the top layers of paper and let the board soak for another minute. Rub the remaining paper off with your fingers. You can also use a soft-bristled toothbrush to remove the paper.

f. Rinse the board with soap and warm water and dry it with a lint-free cloth. If any part of the pattern has rubbed off — especially the circuit portion — use a black permanent marker to make any necessary corrections.

3. ETCH AND DRILL THE BOARD

a. Place the etchant in a plastic tray, and place the board in the etchant solution. Wearing rubber or latex gloves, continually agitate the etchant solution. Once etching is complete (when the copper is completely removed from the non-black areas), remove and wash the board with soap and warm water. Dry the board with a lint-free cloth.

b. Using the steel wool, scrub the board until all the toner is removed. Then rinse the board and dry it with a lint-free cloth.

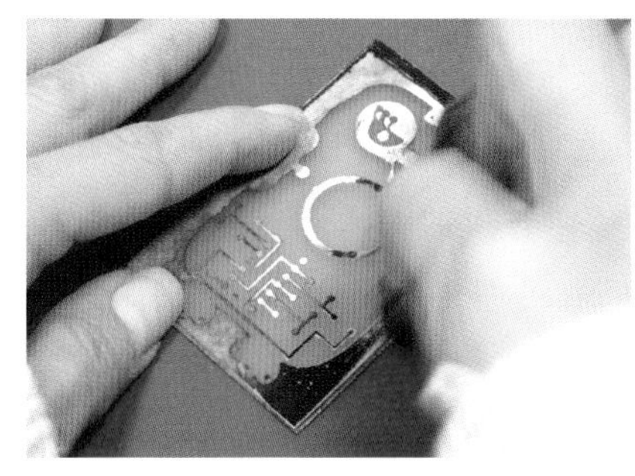

c. Clamp the etched board onto a piece of wood. Using a hand drill and 1mm drill bit, drill holes in the center of each circle of the circuit layout.

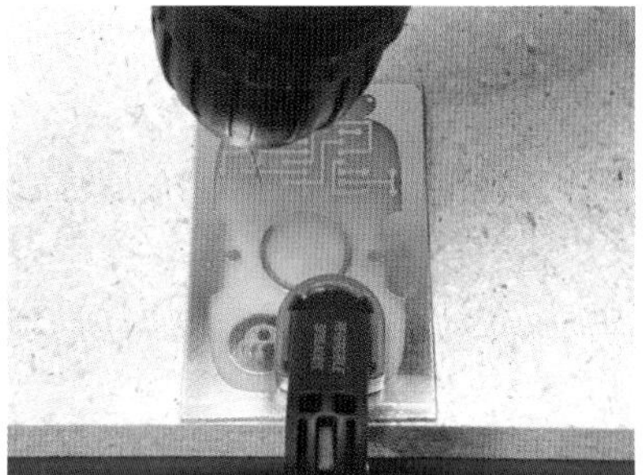

⚠ SAFETY WHEN USING ETCHANT

- **ALWAYS USE ETCHANT IN A WELL-VENTILATED AREA, PREFERABLY OUTSIDE.**
- **WEAR RUBBER OR LATEX GLOVES WHEN HANDLING ETCHANT.**
- **NEVER PLACE ETCHANT IN A METALLIC CONTAINER. Always use a plastic container and plastic tongs.**
- **DO NOT DISPOSE OF ETCHANT DOWN ANY DRAIN. The etchant removes copper and can damage any copper pipes. Dispose of etchant solution properly, according to the instructions provided on the package.**

BREADBOARD PROTOTYPE

STEP 1

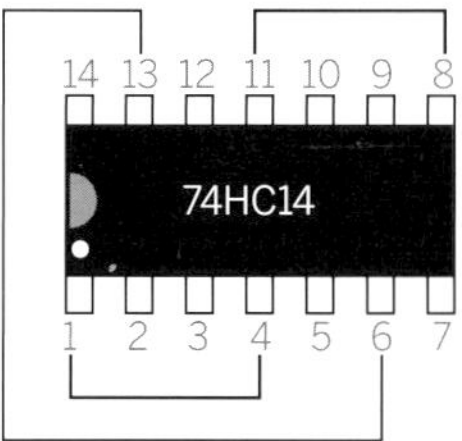

- Connect pin 1 to pin 4
- Connect pin 8 to pin 11
- Connect pin 13 to pin 6

STEP 2

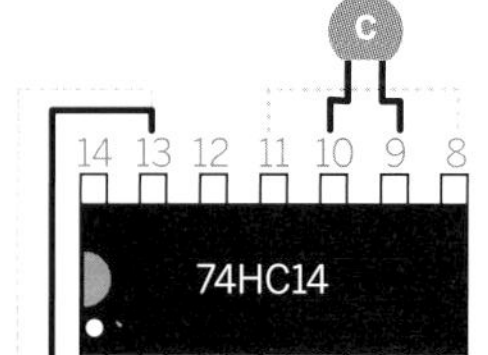

- Connect capacitor A from pin 2 to pin 3
- Connect capacitor B from pin 5 to pin 13
- Connect capacitor C from pin 9 to pin 10

Capacitors A–C can range in value from 1μF to 47μF, and can be ceramic or electrolytic.

STEP 3

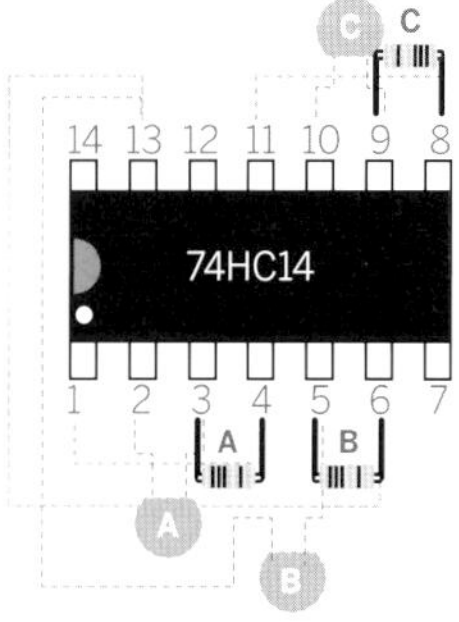

- Connect resistor A from pin 3 to pin 4
- Connect resistor B from pin 5 to pin 6
- Connect resistor C from pin 8 to pin 9

Resistors A–C can range in value from 1K to 100K.

STEP 4

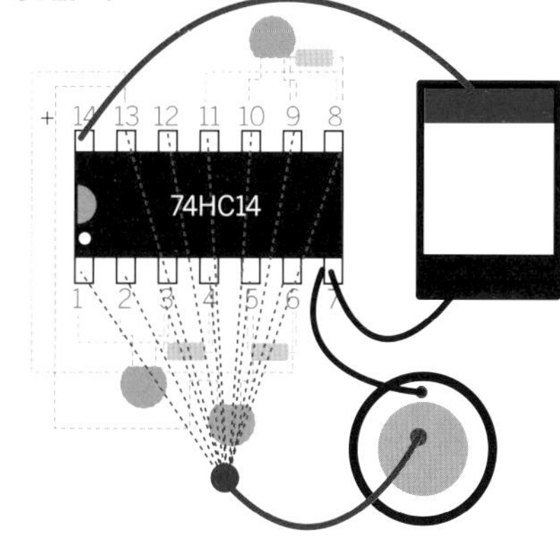

- Connect the positive solar panel lead to pin 14
- Connect the negative solar panel lead to pin 7
- Connect the positive speaker lead to any of the pins EXCEPT pin 7 and 14
- Connect the negative speaker lead to pin 7

4. SOLDER THE COMPONENTS

For best results, build the circuit on a breadboard before you solder the components onto the etched board in order to achieve the best sound. Experiment with the circuit by swapping different value capacitors and resistors to create varying sounds.

a. Grab the IC 74HC14 chip and place its pins accordingly in the appropriate drilled holes with the notch facing Aerial's feet. You may need to gently bend the leads inward to get it to fit comfortably in the drilled holes. Push the pins *halfway* through the drilled holes. You will be soldering on top of the board.

BIRDIE CIRCUIT

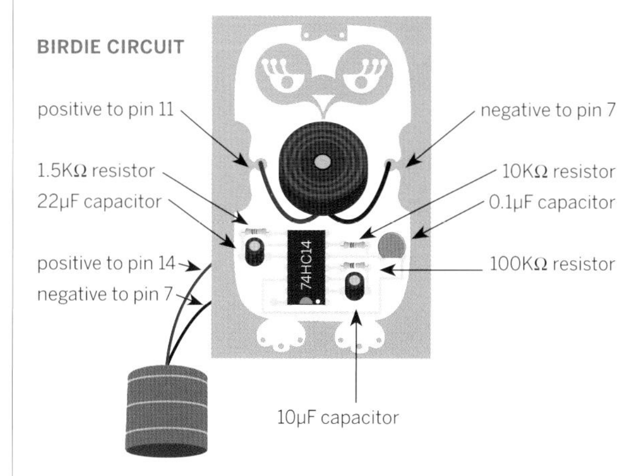

NOTE: Make sure that the notch (half circle) on the IC matches the notch on the circuit layout; otherwise, the circuit will not work properly. The notch should be facing Aerial's feet.

b. Solder the pins onto the board, making sure that each pin makes a solid connection accordingly with its conductive copper trace.

c. Take the piezo speaker and slip the leads through the drilled holes onto the back of the etched board. Slip the stripped end of the negative lead through the hole leading to pin 7. *Do NOT solder the negative lead at this time.* Slip the stripped end of the positive lead through the hole leading to pin 11 (or to the desired pin location if you are creating your own sound). Solder the positive speaker lead only onto the board, making sure that the pin makes a solid connection with its conductive copper trace.

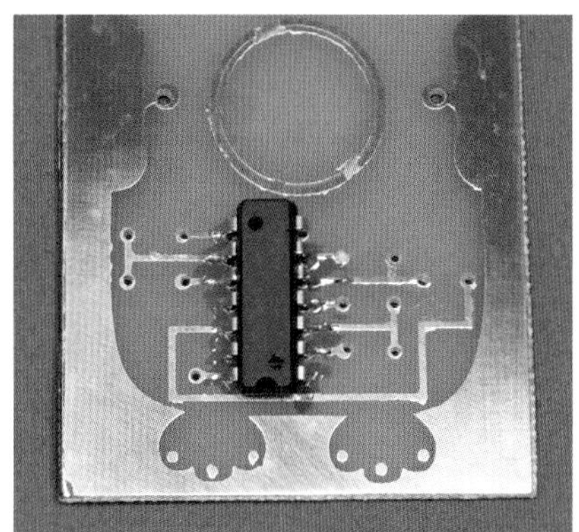

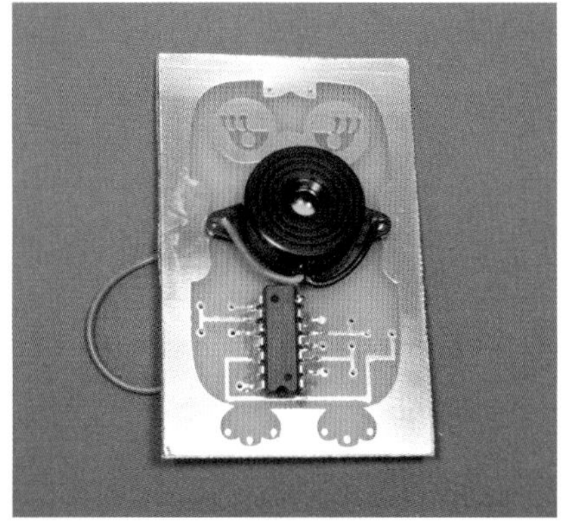

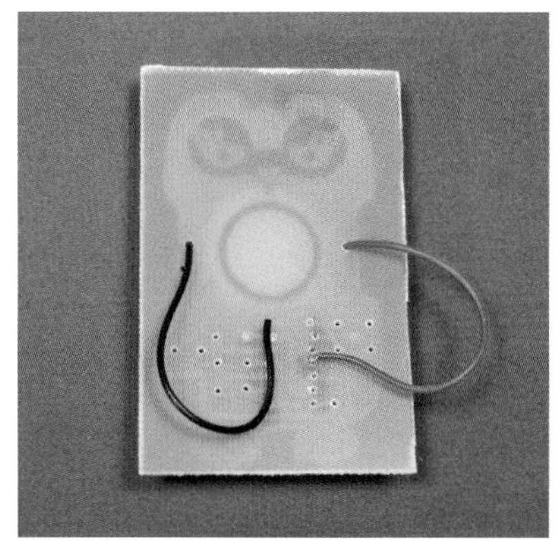

d. Take capacitor A (10μF) and slip its leads through the holes leading from pin 2 and pin 3. Leave enough of the leads above the board so that you can easily solder it onto the conductive traces. Bend its leads in opposite directions on the back of the board to temporarily hold it in place. Solder the leads according to the conductive traces. Then repeat, soldering resistor A (100KΩ) to the conductive traces leading from pin 3 and pin 4. Polarity does not matter here.

e. Take capacitor B (0.1μF) and solder it to the conductive traces leading from pin 5 and pin 12. Then solder resistor B (10KΩ) to the conductive traces leading from pin 5 and pin 6. Solder the last capacitor, C (22μF), to the conductive traces leading from pin 9 and pin 10. Then solder resistor C (1.5KΩ) to the conductive traces leading from pin 8 and pin 9.

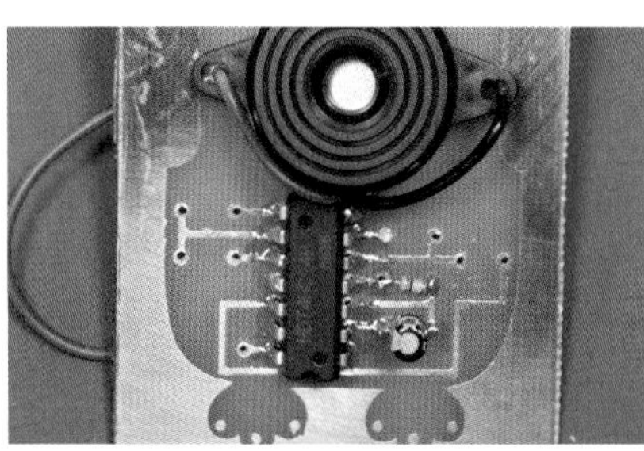

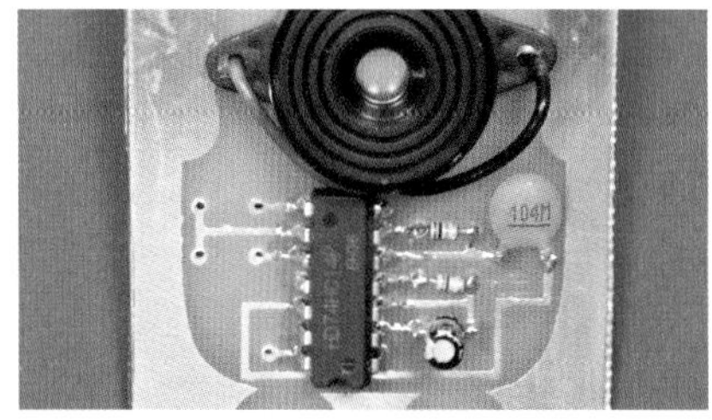

Note: In this particular circuit design, if you are using an electrolytic (polarized) capacitor, the polarity is insignificant because of the low energy flow.

f. Using wire cutters, cut the 8" wire into two 4" pieces. Using wire strippers, strip about ¼" off of both ends. Solder the wire to the contact points on the solar panel. Using a red marker, mark the positive contact. If you don't know which is the positive contact, use a multimeter to distinguish the positive from the negative. (Refer to page 71 to learn how to use a multimeter.)

g. To give the wires some support and integrity at the connection point to the solar cell, you will need to add 1" of heat-shrink tubing. Using scissors, cut the heat-shrink tubing into two 1" pieces. Slip the heat-shrink tubing over each wire, and using a heat gun or hair dryer, shrink the tubing until snug.

h. Roll the solar panel into a cylinder, and using double-sided tape, secure the overlapping edges in place.

i. Slip the positive lead of the solar panel from the back of the etched board through the hole leading from pin 14. Solder. Repeat for the negative lead, slipping the lead through the hole from pin 7 and soldering it in place. Turn the board over and attach the brooch pin.

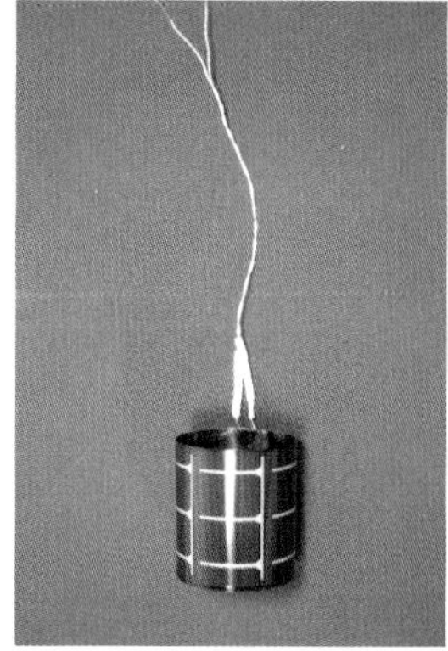

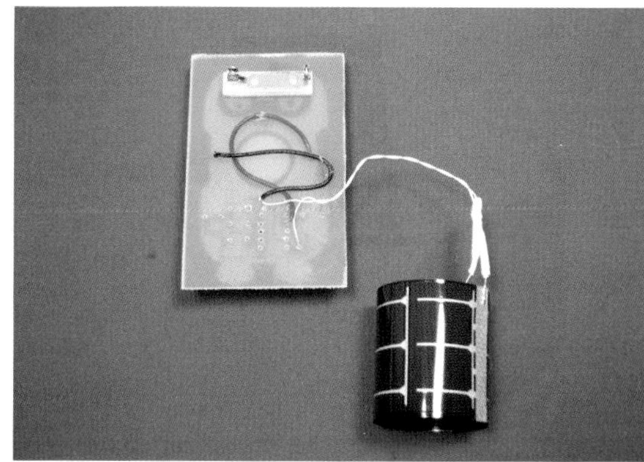

NOTE: The solar panel dangles so it can move around. The movement helps create different sounds.

Clip Aerial onto your lapel or messenger bag, and stroll through the urban streets accompanied by her modest, songful voice.

FINISH 

4

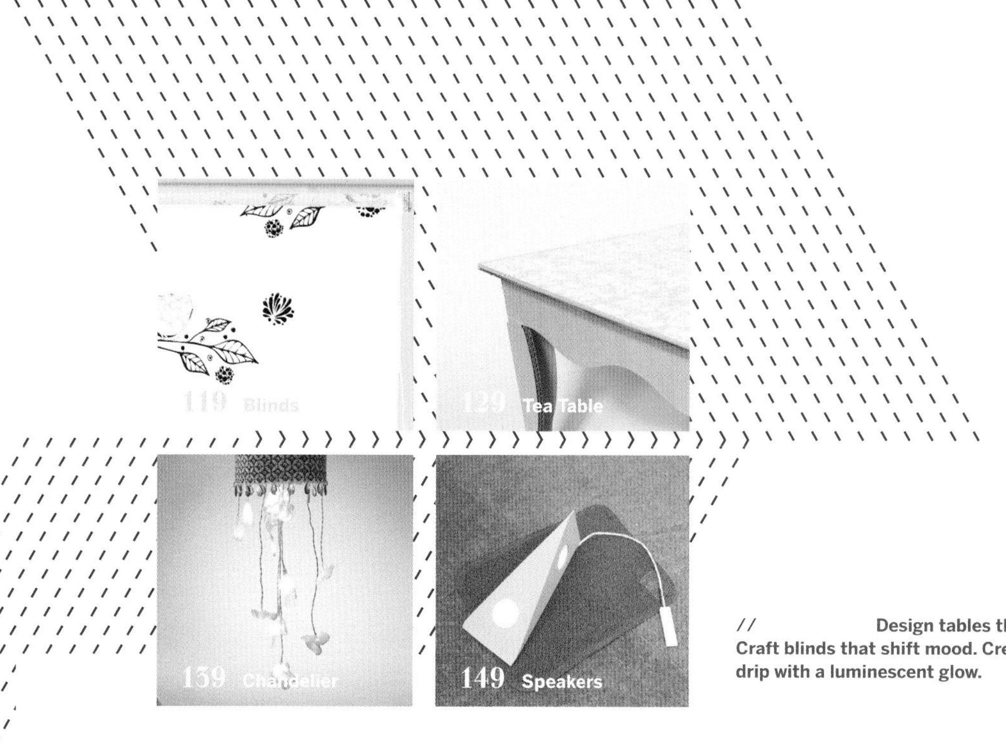

// Design tables that light themselves. Craft blinds that shift mood. Create flowers that drip with a luminescent glow.

HOME ACCENTS

A new generation of glowing and color-changing inks, LEDs, and smart plastics offer you the opportunity to craft high-tech interiors that have charisma and poetic life. Subtly transform your home into a fantastical realm, decorated with dynamic and radiant skins that shift patterns and moods from dusk to dawn. Exploring novel and lyrical ways to interact with technology, you can even hand-craft electronic gadgets and lighting fixtures without a single toggle switch in sight. The tutorials in this chapter introduce you to crafting home furnishings and gadgets that are alive with a sense of functionality and theatrical flair.

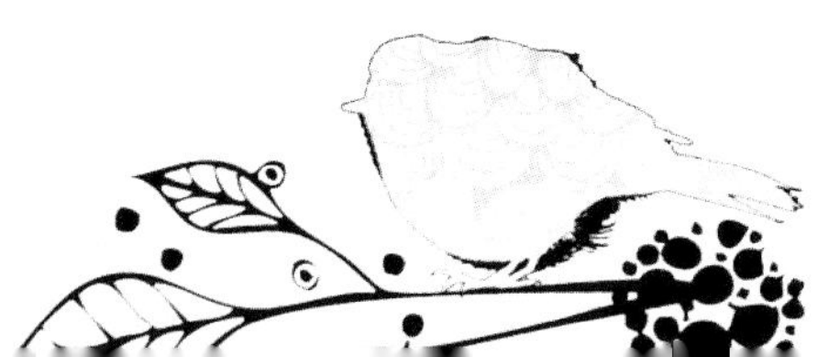

PHOTOCHROMIC BLINDS

Animate your interiors with textiles that dynamically blossom into color during the day and secretly fade at night. Using traditional screen-printing processes, decorative patterns can be printed onto textiles, such as blinds. Supplementing conventional inks with photochromic inks creates patterns that appear and then disappear when a UV light source, such as the sun, is removed.

HOW IT WORKS »

Photochromic inks change from clear to a color when excited by UV light (sunlight). The UV light triggers the molecules of the photochromic ink to temporarily change, resulting in a burst of color. The color change will occur over and over.

COMPLEXITY: BASIC

WHAT YOU WILL LEARN

» Screen-print with photochromic ink

RELATED TUTORIALS

» The Basics of Screen Printing (51-55)

MATERIALS

» **Roller blinds** with white fabric
» **Photochromic ink**
» **Fabric ink**
» **11"×17" transparency film (4)**
» **Masking tape**
» **Screen-printing supplies** Refer to page 52 for details.
» **Screen with mesh between 85–110** for photochromic image
» **Screen with mesh between 110–150** for fabric ink image

TOOLS

» **Plastic mixing knife**
» **Scissors**
» **Heat gun or blow dryer**
» **Computer and printer**
» **Masking tape**
» **Cellophane tape**

PATTERNS

» **Pattern A**
Branch and bird outline
» **Pattern B**
Bird filler
» **Pattern C**
Registration mark

PATTERNS

Patterns available at fashioningtechnology.com/blinds.

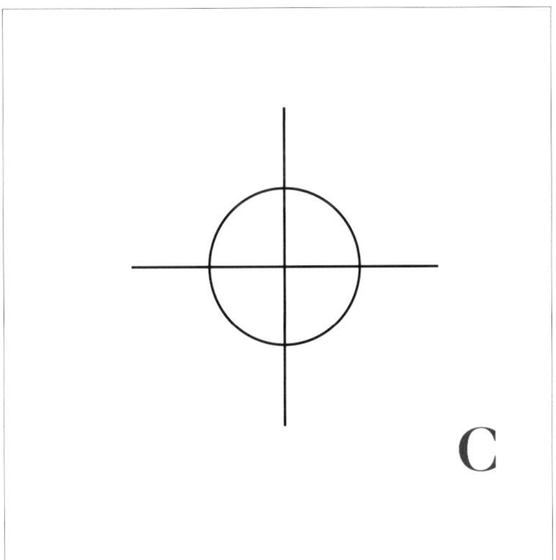

START »

1. SELECT YOUR BLINDS

To make the Photochromic Blinds, you'll need to either purchase a set of roller blinds with white fabric or replace the fabric of another pair of blinds with the fabric that you will screen-print. Measure your windows and purchase a set of blinds that works for those measurements. Remember that the fabric *must* be white for the photochromic ink to work.

2. MAKE THE SCREENS

You will need 2 screens with different screen meshes. The screen to be used with the photochromic ink needs to be a coarser screen with a mesh between 85–110. The screen for the fabric ink can have a finer mesh between 110–150. To learn more about screen meshes, refer to page 52.

The size of your screen must be slighter larger than your graphics. The following tutorial uses a 18"×22" screen burned with 2 images, on opposite sides, to expedite the printing process. If your blinds are fairly wide, print the pattern in the center of the blinds. Another option is to make an extra set of screens (for the second set of images that are mirrored) and print each set of patterns, one at a time.

NOTE : **The screens with Patterns A and B both contain registration marks that help you align the 2 prints. If you are creating your own patterns, it would be helpful to include registration marks, but it's not absolutely necessary.**

? What are registration marks?

» Registration marks are often used in screen printing to help align graphics when more than one ink is used. Typically, the same mark is burned at the exact same location on same-sized screens, away from the graphic that is to be transferred.

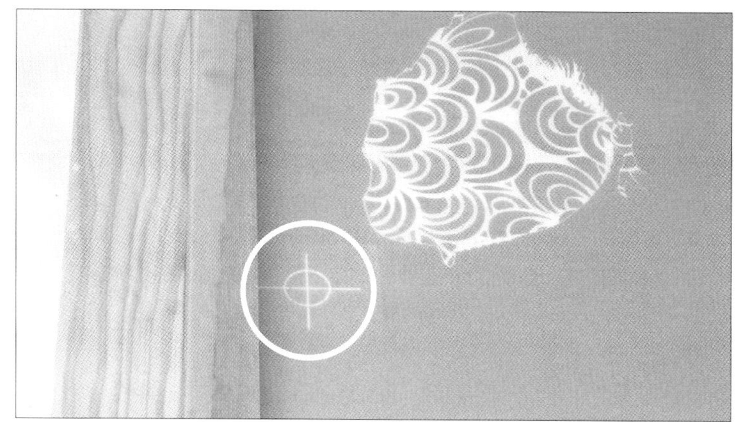

a. Print 2 copies of Pattern A on 11"×17" transparencies. Print 4 registration marks on transparency film. Cut out the registration marks. Using cellophane tape, place one of the registration marks near the center, top edge of Pattern A. Flip the second transparency of Pattern A horizontally. Using cellophane tape, place another registration mark near the center, bottom edge.

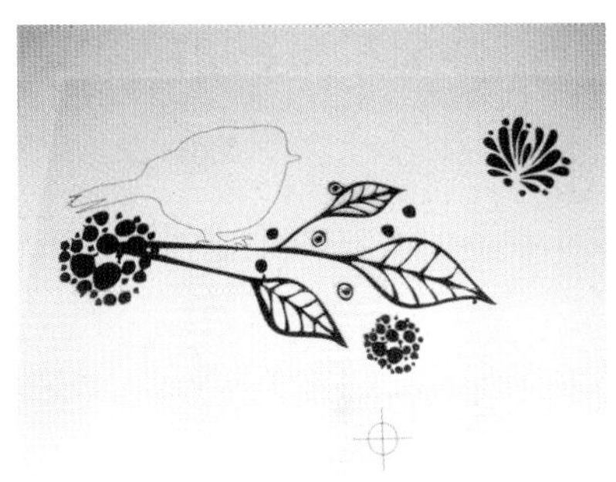

b. Print 2 copies of Pattern B on 11"×17" transparencies. Place one of the transparencies of Pattern B directly on top of one of the transparencies of Pattern A, aligning the 2 images. Once the images are aligned, place the third registration mark on Pattern B so that it lines up with the one on Pattern A. Repeat for the second transparency for Pattern B, adding the fourth registration mark.

c. Once you have coated the screen with photo emulsion and the screen has dried, place the transparencies on separate screens so that the images align with each other. Using cellophane tape, secure the transparencies for Pattern A on the finer mesh screen, taping one near the upper edge and the other near the bottom edge of the screen. Repeat for Pattern B on the coarser screen, making sure that placement of the images on both screens line up with each other.

d. Following the instructions on pages 54–55, burn Pattern B on the coarser screen and Pattern A on the finer screen. You can also create your own screen-printing patterns.

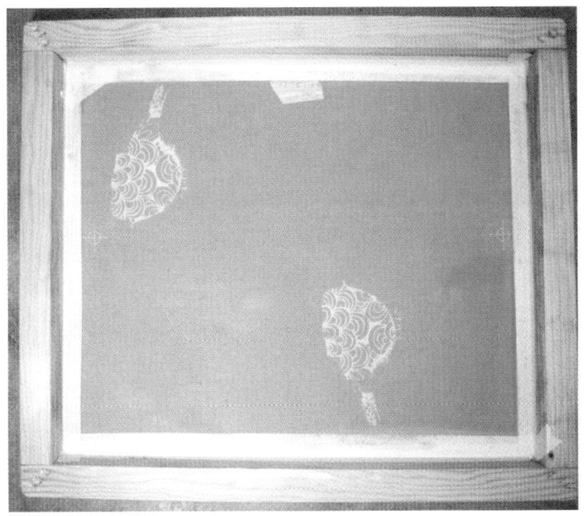

3. PRINT THE PHOTOCHROMIC LAYER

a. Place the fabric on a sturdy, flat surface. Using masking tape, tape the edges of the fabric to the table. This will ensure that the fabric doesn't move in the printing process. Consider laying down a piece of scrap paper beneath the fabric to protect your surface in case the ink bleeds through.

b. Position the transparency for Pattern B 3"–4" from the top and 1" from the edge of the fabric. If you are using registration marks to align the patterns, tape down a thick piece of paper on the fabric where the registration marks will be printed. You want the registration marks to print on the paper, not the fabric. Place the screen directly on top of the transparency, aligning the burned image with the black patterns on the transparency.

c. Using a plastic knife, mix the photochromic ink thoroughly. Place a bead of photochromic ink at the side of each image and registration mark. Lift the screen about 2" off the fabric and, using a squeegee, flood the image with the ink, moving the ink away from you.

NOTE: To get even ink coverage, your squeegee should be slightly larger than your image area.

d. Carefully lower the screen. Hold the squeegee at a 45° angle at one side of the screen. Applying even, firm pressure, move the ink toward you, printing the image. Repeat for the pattern on the opposite side. Lift the screen. Let the image dry before you make the next print.

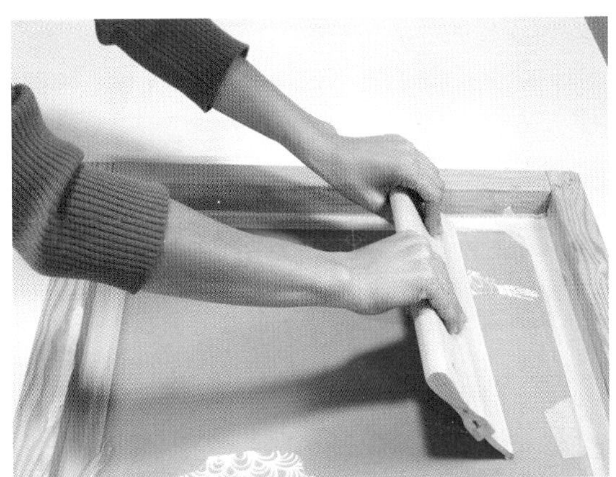

TIP: You can use a hair dryer or a heat gun set to low to expedite the drying process, but be careful not to get the heat too close to your screen, as you could damage it.

e. Clean the screen and squeegee using the appropriate solvent for the ink, otherwise the ink will dry in the screen before you make another print. Let the screen dry before you make the next print.

4. PRINT THE FABRIC INK LAYER

a. Once the image area is dry, you are ready to make the next print. Align the screen burned with Pattern A directly on top of the photochromic image, using the registration marks as a guide. Repeat the process, flooding the screen with fabric ink and making the top print.

b. Lift the screen and carefully remove the registration prints. Clean the screen and set it aside to dry. Let the image dry before you make the next print.

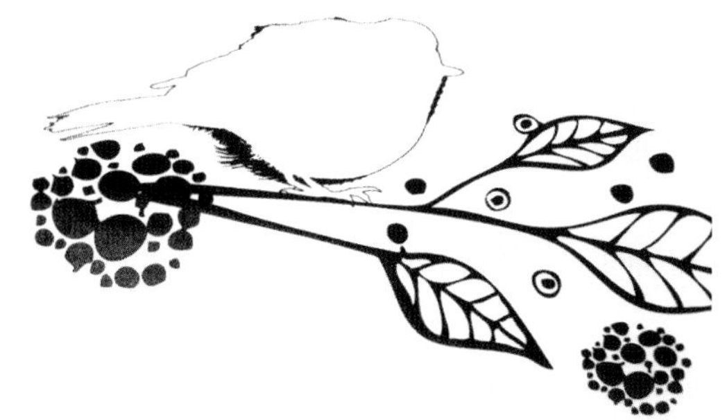

c. Repeat the printing process moving down the fabric, first printing the photochromic layer and then the fabric ink layer until you reach the end of the fabric. Ideally, you should space the pattern out evenly.

5. ASSEMBLE

The final assembly will vary, depending on the roller blinds that you have purchased. If you are not using the fabric that came with the blinds you purchased, remove the old fabric and replace it with the one that you have screen-printed.

FINISH

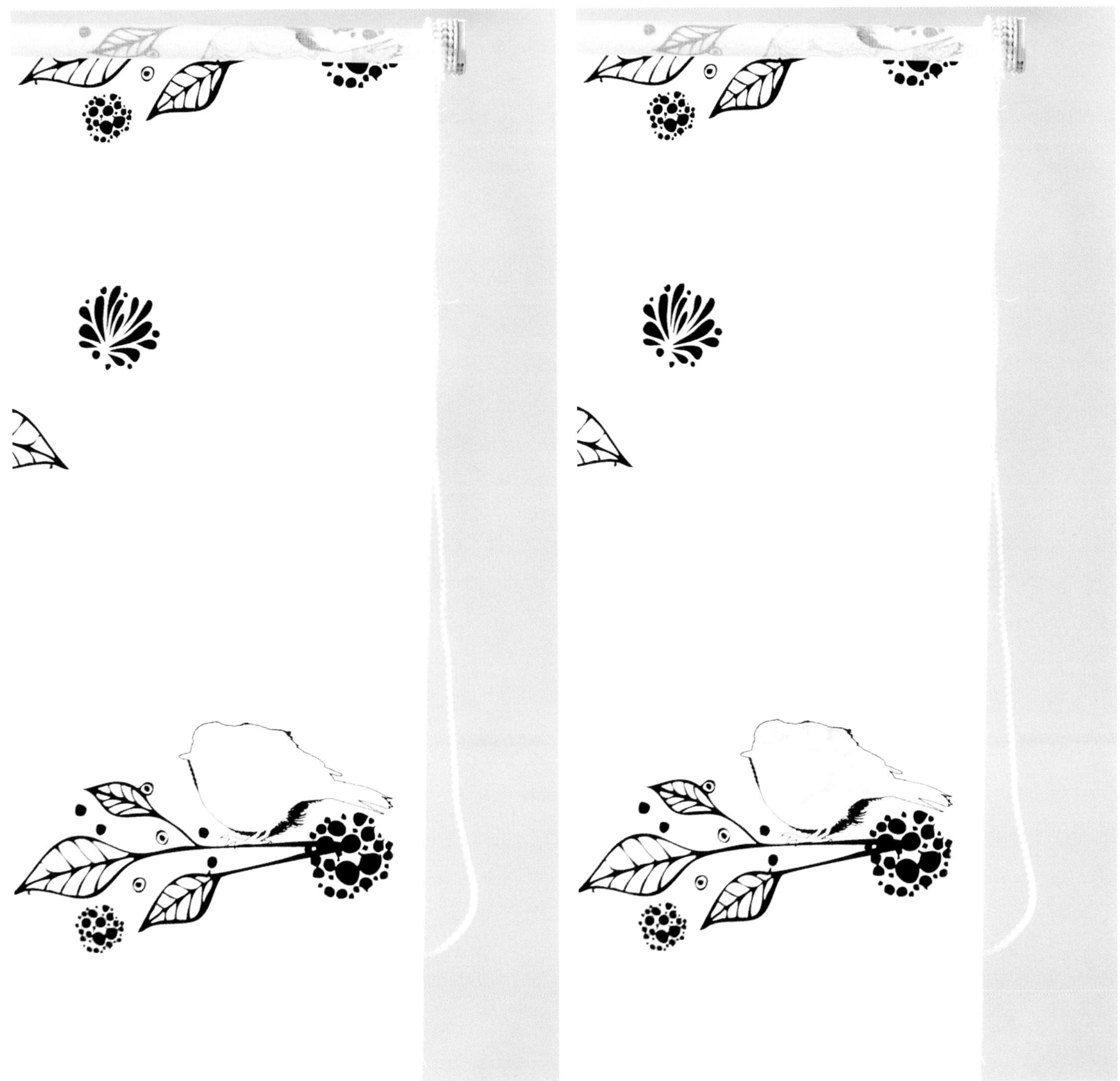

LUMINESCENT TEA TABLE

Put away your candles and instead fashion your furniture to radiate a soft glow. The Luminescent Tea Table features a decorative pattern coated with phosphorescent ink. The pattern absorbs sunshine during the day and emits light at night. The Tea Table doesn't require any electricity and can glow for up to several hours.

HOW IT WORKS »

The Luminescent Tea Table uses phosphorescent inks that have the remarkable ability to absorb and capture UV and artificial light during the day and gradually release this energy in the dark. The duration of the glow depends on two factors: the ink quality and exposure time to light. Phosphorescent inks are nontoxic and free from radioactive additives.

COMPLEXITY: INTERMEDIATE

WHAT YOU WILL LEARN

- Screen-print with phosphorescent inks
- Construct 3D forms from 2D planes

RELATED TUTORIALS

- The Basics of Screen Printing (51-55)

MATERIALS

- **¼" cardboard, 4'×4' (2)**
- **18"×28" brown chipboard** for tabletop
- **18"×28" plexiglass, ⅛" thick**
- **Phosphorescent pigment or ink**
- **Transparent poster ink** only if using pigment
- **White poster ink**
- **Wood glue or all-purpose glue**
- **11"×17" transparency film**
- **Masking tape**
- **Screen-printing supplies** Refer to page 52 for details.
- **Minimum 18"×22" screen with mesh between 85–110**
- **Plastic container**
- **Plastic mixing knife**
- **Rubber band**
- **Clothespins** (optional)

TEMPLATES & PATTERN

- Template A
  Table legs, short (4)
- Template B
  Table legs, long (4)
- Template C
  Tabletop (1)
- Pattern D

TOOLS

- **Ruler**
- **Cutting mat**
- **Pencil**
- **Utility knife**
- **Heat gun or blow dryer**
- **Computer and printer**
- **Masking tape**

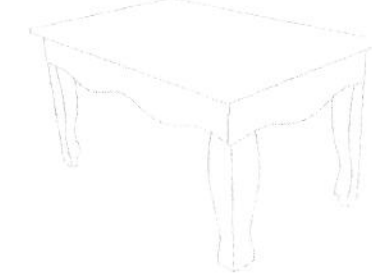

TEMPLATES

Template and pattern available online at fashioningtechnology.com/teatable.

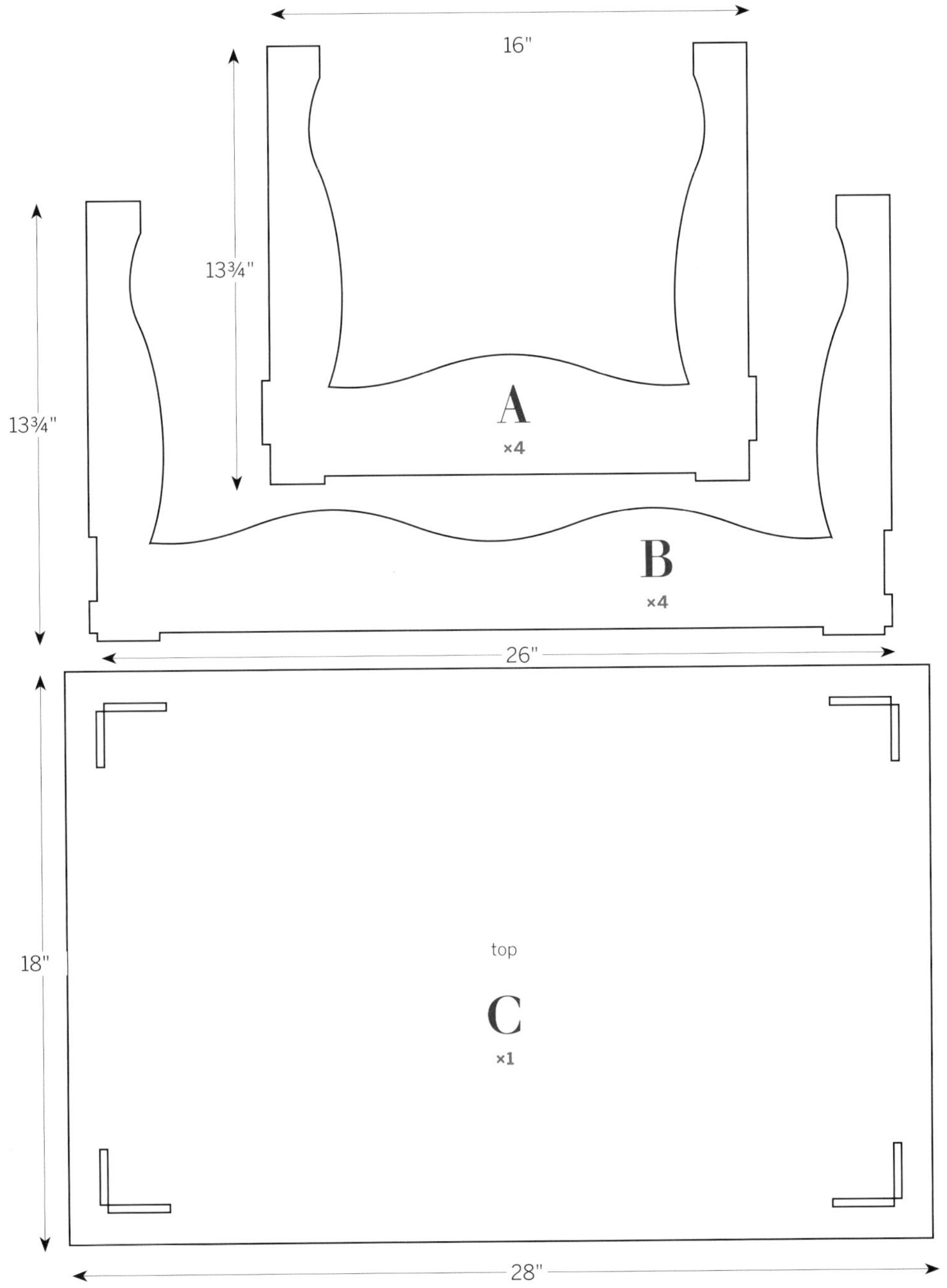

16"

13¾"

13¾"

A

×4

B

×4

26"

18"

top

C

×1

28"

D

START »

1. CUT THE CARDBOARD FORM

a. Using a pencil, trace Template A onto the cardboard. Then, using a ruler and utility knife, cut the template from the cardboard. Repeat for all 3 templates. Remember to cut out the slots for the table leg joints on Template C.

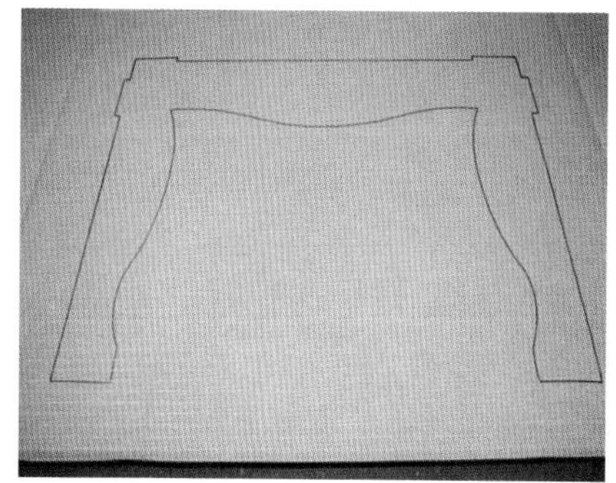

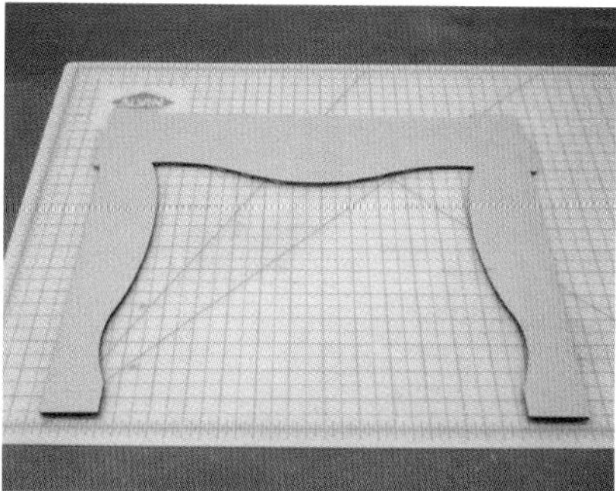

✻ TIP: Don't try to cut through the cardboard in one cut. To get clean cuts, use a fresh blade on your utility knife and a ruler as a guide. You will need to make several, slow cuts. Changing your blade after cutting each pattern greatly helps.

b. Place one of the pieces of Template A on a table. Add a line of glue along the center and, using a scrap piece of cardboard, spread the glue along the cardboard. Place the second piece of Template A directly on top, carefully aligning the edges. Set aside to let the glue dry. Repeat, creating the second set of short table legs for Template A. Repeat the entire process for Template B, creating a pair of double-thick cardboard table legs.

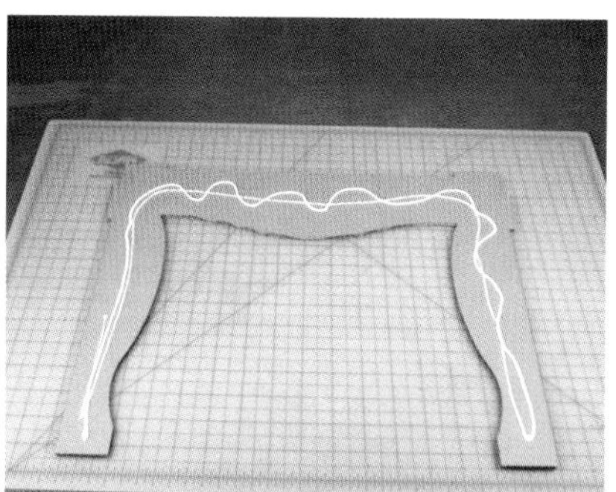

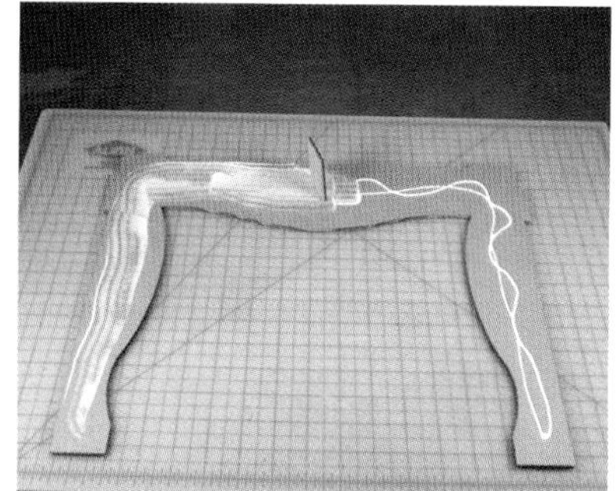

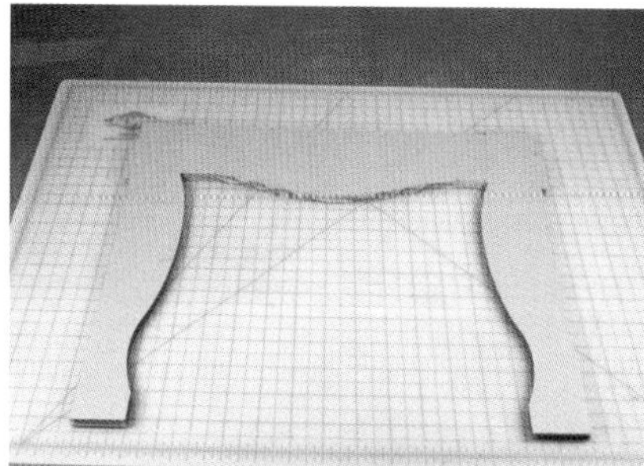

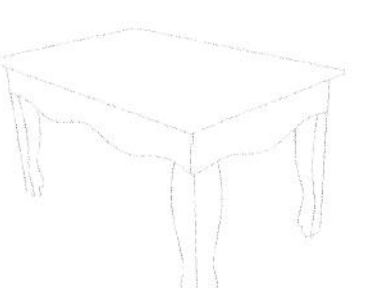

2. CONSTRUCT THE TABLE

a. Place Template C on a table. Slip the joint edges of Template A and Template B into their respective slots in Template C. Add a touch of glue to the joint areas and the inner seam of the legs. You can use rubber bands to temporarily secure the cardboard pieces together while the glue dries.

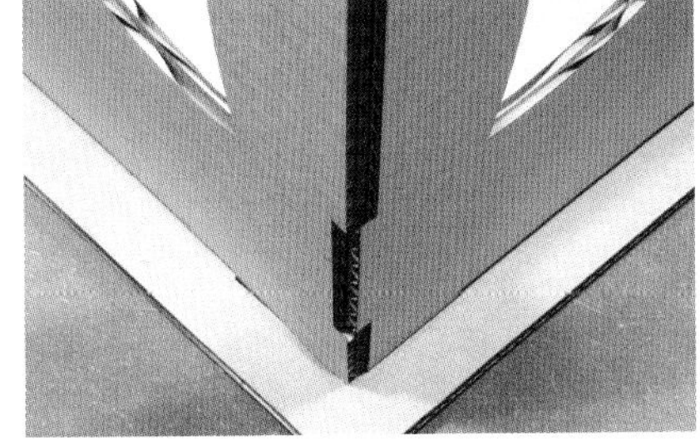

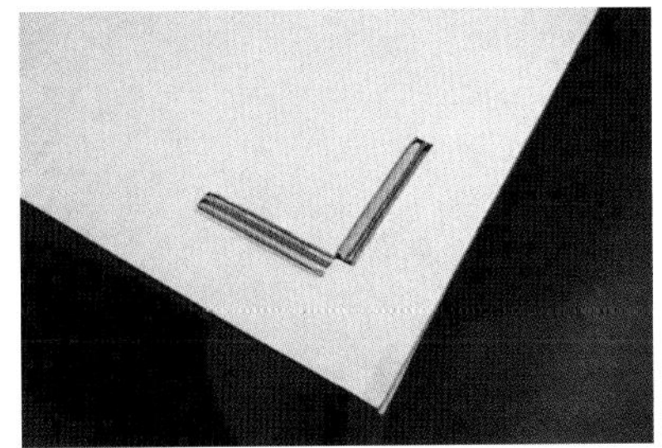

b. Using a utility knife, cut four 2"×1" rectangular pieces of cardboard. These pieces will function as reinforcement tabs for the table legs. Using a utility knife, score all 4 pieces in the center. Add a bit of glue to one side of the reinforcement tab. Adhere a tab at the bottom of each leg.

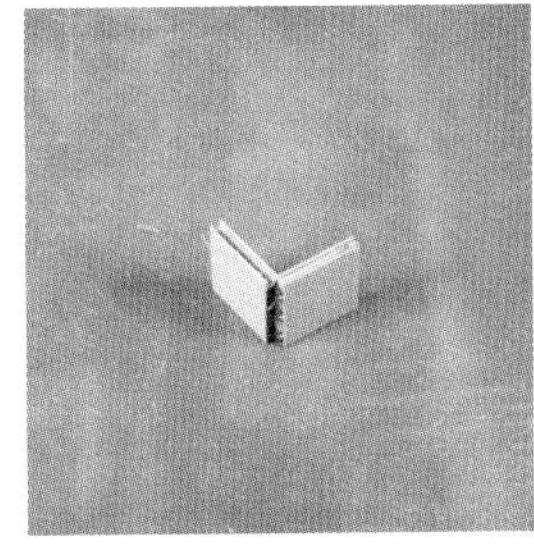

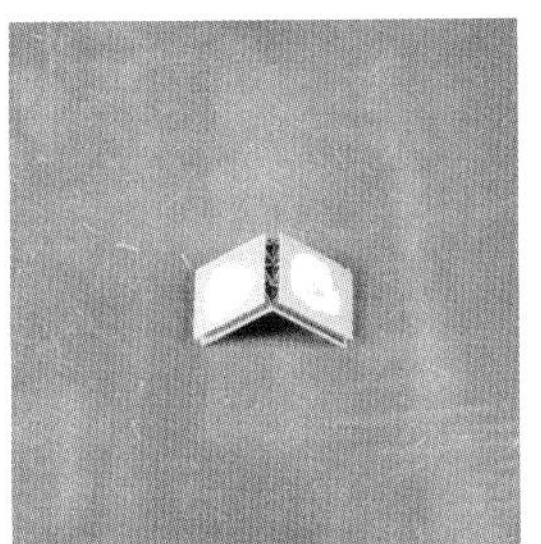

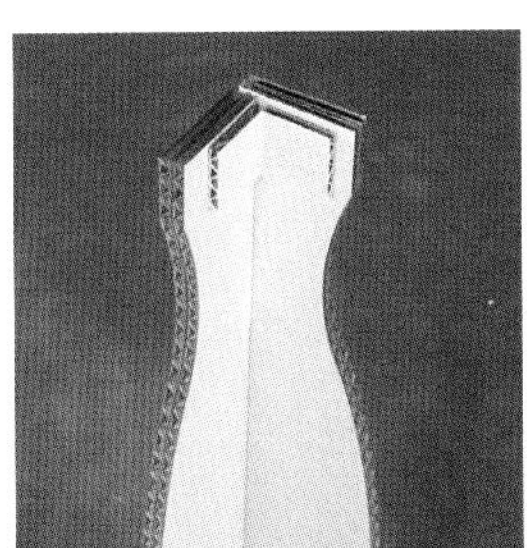

3. SCREEN-PRINT THE TABLETOP

NOTE: Unless you have a more sophisticated screen-printing setup with hinge clamps, it's helpful to have an extra set of hands when working on printing the tabletop.

a. Using a utility blade, cut the chipboard to an 18"×28" rectangle. Following the instructions on page 54, burn the screen with your pattern. You can either download the pattern shown here from fashioningtechnology.com/teatable or create your own.

b. Using masking tape, tape the corners of the chipboard onto the table. This will ensure that the board doesn't move in the screen-printing process. Leaving a ¼" border, line up the screen where you want to make the first print. Using spare pieces of chipboard or cardboard, tape down the board flush to the top and side edges of the top of the frame. These guides will ensure that screen remains in place for printing.

NOTE : If you are designing your own pattern, fine lines and details in your pattern are NOT recommended.

Special thanks to Sara Schmidt for the lovely design of the tabletop pattern.

c. Lift the screen about 2" off the chipboard. Place a bead of white ink at the bottom of the screen and, using a squeegee, flood the image with the white ink, moving the ink away from you.

NOTE: To get even ink coverage, your squeegee should be slightly larger than your image area.

d. Carefully lower the screen. Place the squeegee at the top of the screen, holding it at a 45° angle. Applying even pressure, print the image, moving the ink toward you. Lift the screen. Let the image dry before you make the next print.

e. Clean the screen, using the appropriate solvent for your ink. Let the screen dry before you make the next print. Once the image area and screen are dry, you are ready to make the next print. You are making several prints because the screen is smaller than the surface area of the table, and you are printing along the tabletop in sections. Align the screen next to the first print with the image areas slightly overlapping. Tape down the cardboard guides along the corner edges of the frame. Repeat the process, flooding the screen with ink and making the second print.

f. Before you make the final print, place a piece of masking tape ¼" along the edge of chipboard (that you have not printed on yet). Don't press too hard on the tape, as you don't want it to adhere permanently to the surface. As the image area for the last print is larger than the print area, the masking tape ensures that you have a ¼" border. Repeat the process to make the third and last print. Then clean the screen and squeegee.

❇ **TIP: You can use a hair dryer or a heat gun set to low to expedite the drying process, but be careful not to damage the screen.**

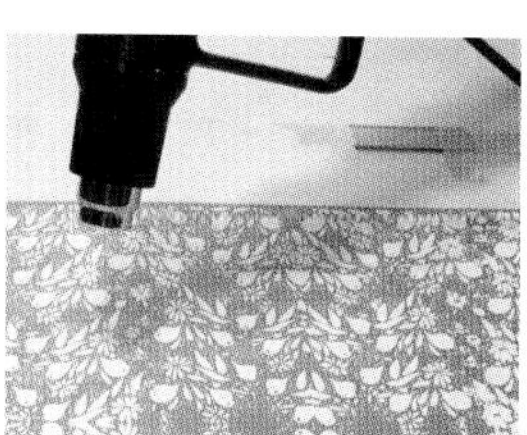

4. PREPARE THE PHOSPHORESCENT INK

a. Fill the plastic container halfway with the transparent base. Add small quantities of phosphorescent powder to the base, thoroughly mixing the 2 together. The amount of powder that you need to add to the base depends entirely on the desired duration and brightness of the glow effect. The more powder you use, the stronger the effect. Keep in mind that the more powder you add to the base, the more the color of the final top coat will be affected.

NOTE: The phosphorescent powder must be mixed with a transparent base. This example uses a bronze liquid with the powder, giving the final application a slightly warmer color.

b. As powders from different manufacturers vary, make test swatches before you apply the ink to the chipboard. To make test swatches, apply several strips of white ink onto a scrap piece of chipboard, and let dry. Then apply a thin coat of the phosphorescent ink mixture directly on top of one of the white ink strips, and let dry. Expose the swatch to UV light (either sunlight or black light) for several minutes. Grab the swatch and take it to a dark environment. If the glow isn't bright enough, add more pigment to the mixture. Keep repeating the process until you achieve the desired level of brightness.

c. Place the screen on top of the chipboard, aligning the screen with the white printed graphics on the chipboard. You want to print directly over the white screened pattern. Using spare pieces of chipboard or cardboard, place the guides flush to the top and side edges of the front of the frame.

d. Place a bead of phosphorescent ink at the bottom of the screen. Lift the screen about 2" off the chipboard and, using a squeegee, flood the image with the ink, moving the ink away from you.

e. Carefully lower the screen. Place the squeegee at the top of the screen, holding it at a 45° angle. Applying even pressure, print the image, moving the ink toward you. Lift the screen. Let the image dry before making the next print. Then clean the screen, using the appropriate solvent for the ink, and let it dry before making the next print.

f. Repeat, making 2 more prints. Don't forget to place the masking tape ¼" along the edge of the chipboard before you make the last print. Let the print dry.

5. ASSEMBLE

a. Place the chipboard face down. Add a bead of glue along all 4 edges. Using a scrap piece of cardboard, spread the glue about 1" along the chipboard edges. You don't need to cover the entire back of the chipboard with glue. Then place the chipboard right side up, directly on top of the tabletop, carefully aligning the edges.

b. Place the 18"×28" plexiglass directly on top of the chipboard, being careful not to move the chipboard out of alignment. If you have clothespins lying around, you can gently clamp the sides of the 2 boards and the plexiglass with pins until the glue dries.

NOTE: Most hardware stores will cut plexiglass to your desired size. You can cut the plexiglass yourself using a utility knife and a ruler to score it, then snap it along the score lines over the edge of a table.

FINISH

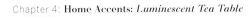

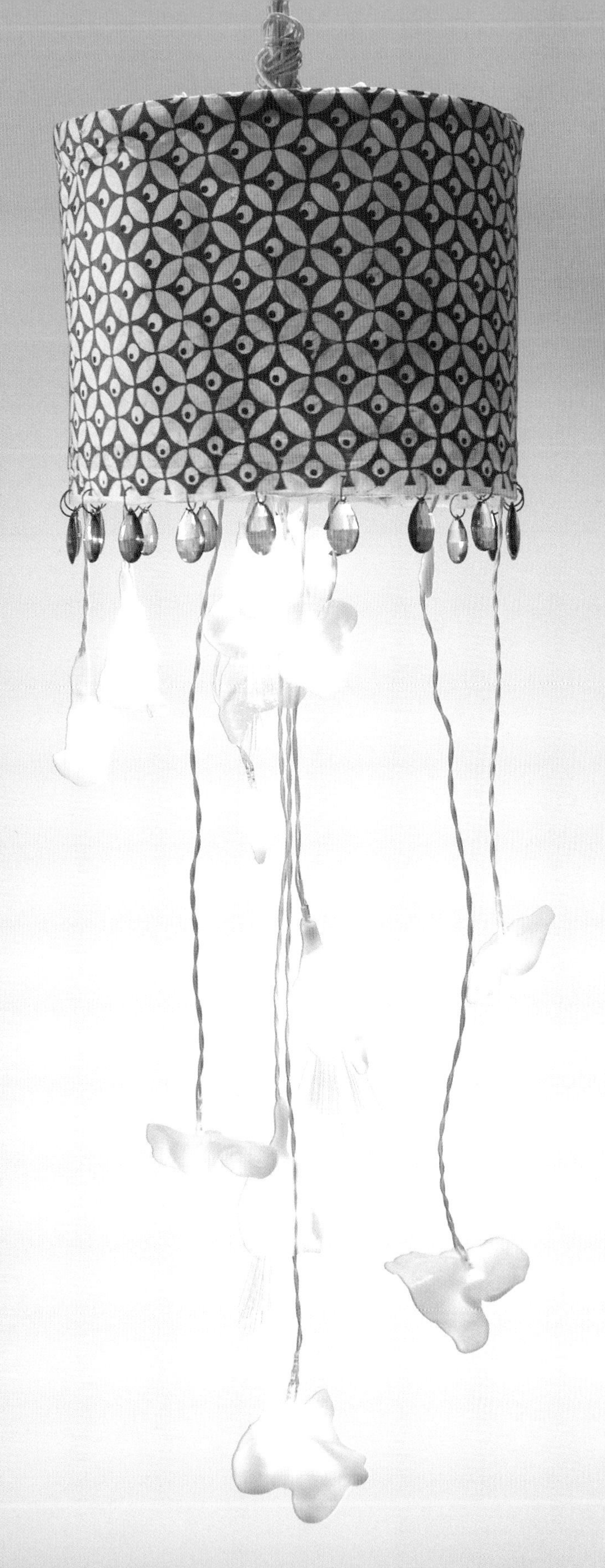

LED CHANDELIER

The LED Chandelier is made from 16 tiny, ultrabright LEDs and Polymorph plastic, hand-sculpted into a variety of floral forms. A vine of LED flowers drips from the chandelier's center, creating a soft, atmospheric mood. A poetic lighting sculpture, the LED Chandelier brings a glitter of glamour to any room.

HOW IT WORKS »

Just like any ordinary lamp, the LED Chandelier is plugged into a wall outlet in order to turn it on. Yet, unlike most typical lamps, the chandelier turns on and off using a novel magnetic switch. Dangling low from the center of the chandelier are two small magnets. When the magnets snap together, the LED flowers begin to glow.

COMPLEXITY: INTERMEDIATE

WHAT YOU WILL LEARN

- Wire multiple LEDs
- Sculpt with Polymorph

RELATED TUTORIALS

- Working with LEDs (19-23)
- Using Resistor Calculators (21)
- Power Adapters (42)
- Hacking into a Power Adapter (43)

MATERIALS

- Polymorph plastic, 200g–500g
- Fiber optics, 12', clear/white
- Spool of 22 AWG stranded wire
- Spool of 24 AWG solid wire
- Conductor white lamp wire, 9'–15'
  Length depends on environment.
- Embroidery hoop, 9" diameter
- 1/8" white heat-shrink tubing, 5'
- 3/16" white heat-shrink tubing, 3½'
- 3/8" white heat-shrink tubing, 1'
- Solder
- Dowels, 9⅝"-long, 3/16"-diameter (2)
- White paint brush-on or spray paint
- Decorative fabric or paint
- Scrap piece of wood for drilling
- Electrical tape
- Chandelier beads
- Bag of silver jewelry jump rings
- Boning, 60" available at fabric stores
- Glue
- Zip tie
- Rubber band

TOOLS

- Clamps
- Multimeter
- Drill and 3/16" drill bit
- Heat gun or hair dryer
- Scissors
- Soldering iron
- Needlenose pliers
- Pencil or marking pen
- Sewing machine
- Wire cutters
- Wire strippers
- Black permanent marker
- Hot glue gun
- Pencil
- Fabric tape ruler
- Plastic or wooden tongs
- Small clips or clothespins (optional)

ELECTRONICS

- 5mm super-bright white LEDs (16)
- Resistors (16)
- Regulated power adapter between 3.7V–9V
- Magnetic switch/connectors
  mutr.co.uk product #EM5 013

START »

1. DETERMINE YOUR RESISTOR VALUE

Before you begin wiring your LEDs, you first need to determine the value of the resistors you need. You will be wiring the LEDs in parallel and adding a resistor to each LED. To determine the resistor value, you need the following information:

1. Voltage from your power source
2. LED forward voltage (V_F)
3. LED forward current (I_F, typically 20mAh)
4. Number of LEDs (16)

Refer to page 21 for detailed instructions on how to use a resistor calculator.

To wire each LED you will need: two 32" strands of white wire stripped about ½" on both ends, one 3" piece of ⅛" heat-shrink tubing, 1 LED, and 1 resistor. For half the LEDs, you should use 22 AWG stranded wire and for the remaining ones, 24 AWG solid wire. The LEDs using the stranded wire will hang nicely from the chandelier and the LEDs with the solid wire can be twisted and shaped like branches to give the chandelier dimension.

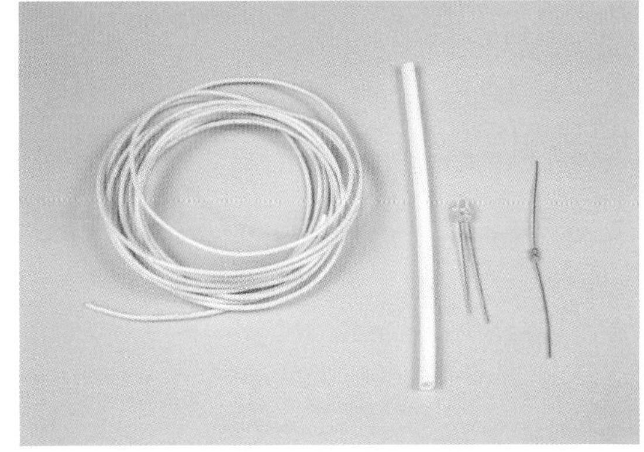

NOTE: To expedite the soldering process, first cut all the materials to their appropriate lengths and set them aside in separate batches.

? Why do I have to add a resistor to each LED rather than adding one resistor for all the LEDs?

» Adding a resistor to each LED allows a greater level of flexibility in your design. You can add additional or fewer LEDs to the chandelier without having to recalculate the resistance you need for the circuit. You can also easily add different-colored LEDs to the circuit, each with its appropriate resistor value. More importantly, adding a resistor to each LED makes the circuit foolproof. For example, if one of the LEDs stops working for some reason, it will not affect the other LEDs in the circuit.

2. WIRE INDIVIDUAL LEDS

a. Gently twist one of the leads of a resistor to the positive (longer lead) of an LED. Using a soldering iron, add a touch of solder, connecting the resistor lead to the LED. Using wire cutters, trim the LED lead so that it doesn't touch the second lead of the resistor. Set aside and repeat for the remaining 15 LEDs.

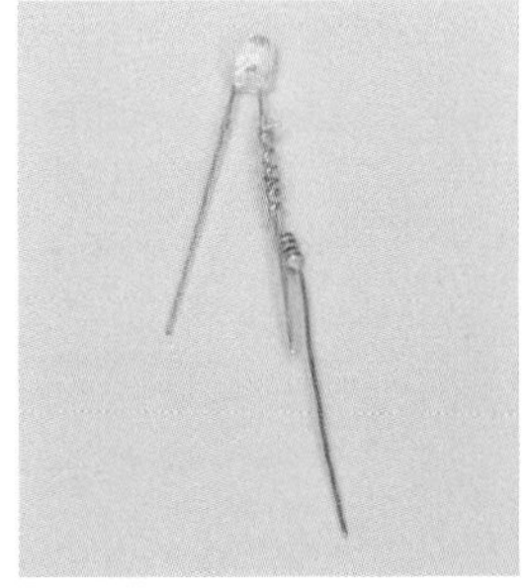

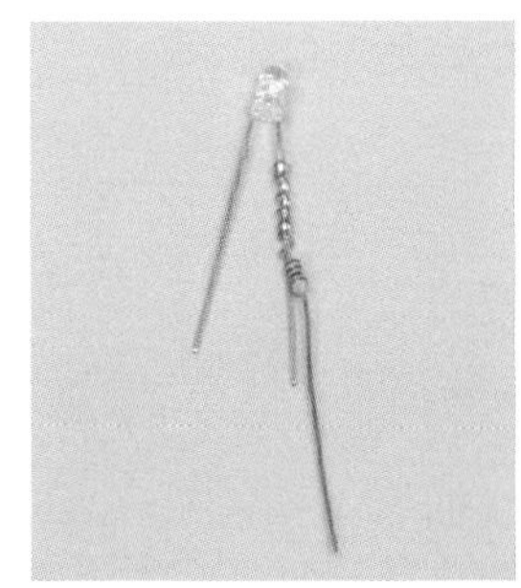

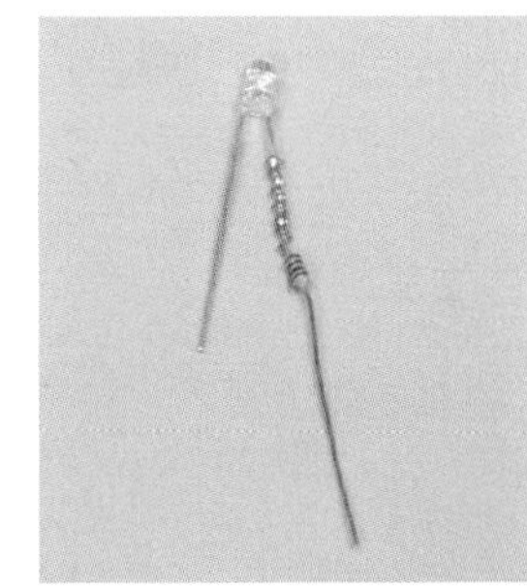

b. Using wire strippers, strip all the wires about ½" on both ends. Twist one of the stripped ends to the second lead of the resistor. Twist a second piece of wire to the negative lead of the LED. Solder. Using a black marker, mark the opposite end of the wire connected to the negative LED lead. This will help you distinguish the positive leads from the negative. Set aside and repeat for the remaining 15 LEDs.

c. Cut the 3" piece of heat-shrink from Step 1 to the appropriate lengths needed to cover the exposed leads. Slip the heat-shrink over an exposed lead and, using a heat gun, shrink the tubing until it is snug. Set aside and repeat for the remaining 15 LEDs.

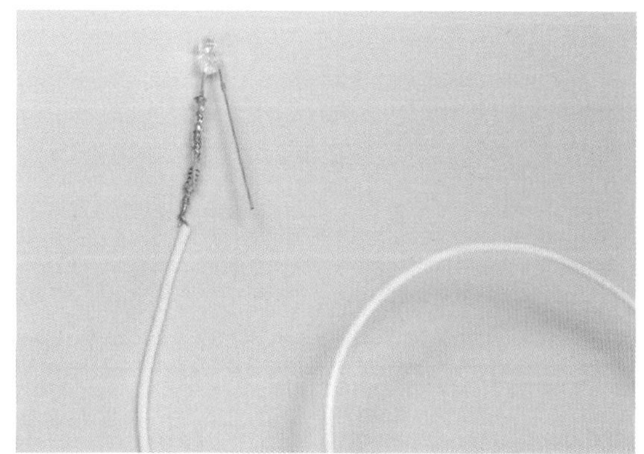

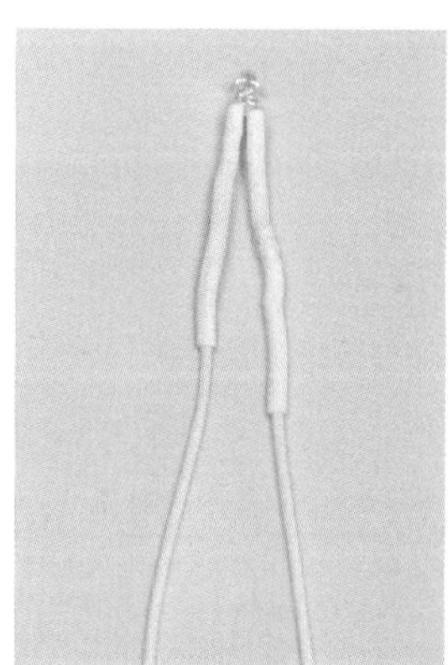

3. CREATE THE MAGNET SWITCH

The LED Chandelier uses a magnetic switch to turn on and off. The magnetic switch is composed of 2 magnets that close the circuit, switching the light on when they are connected.

a. Using wire cutters, cut 2 pieces of 22 AWG stranded wire 48" in length. Using wire strippers, strip about ½" off each end. Wrap one of the stripped ends of wire around the lead of one of the magnet switches. Repeat for the second magnet switch. Solder the wire to the leads.

b. Cut a piece of ⅛" heat-shrink tubing and slip it over the bare metal. Using a heat gun, shrink the tubing until it's snug. Repeat for the second magnet switch.

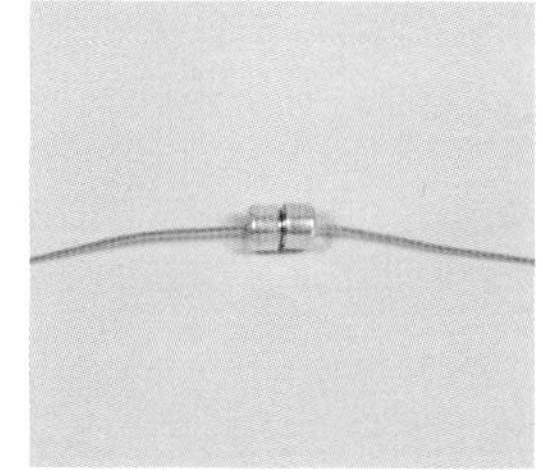

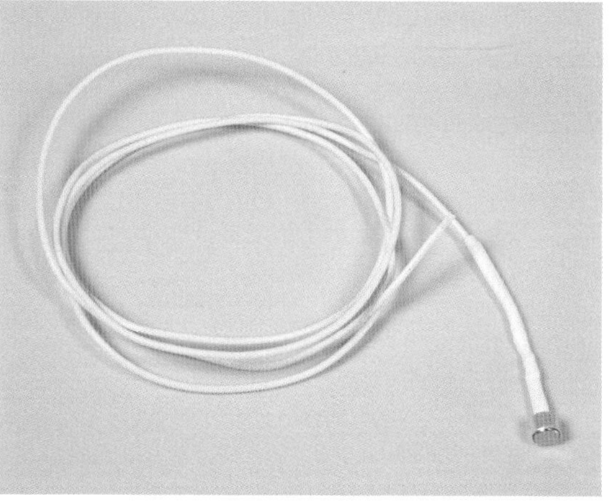

4. ADD FIBER OPTICS TO LEDS

For a few of your LEDs, you may want to add fiber optics and give the overall design a little variety.

Cut the fiber optics to about 2" in length. Secure the cluster together using a rubber band at one end. Using a hot glue gun, add a touch of hot glue to the top of an LED. Quickly place the fiber optics at the top of the LED, holding the bundle in place for 10–15 seconds or until the fiber optics adhere to the LED casing. Make a few more.

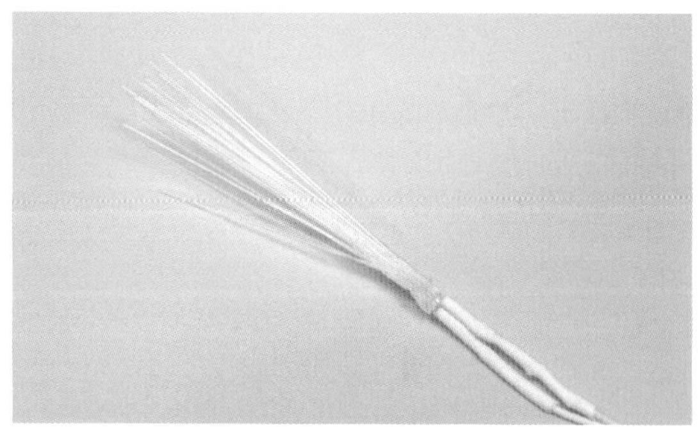

5. WORK THE POLYMORPH

You will be using Polymorph plastic to hand-sculpt the floral forms around each LED. Heat the Polymorph according to the instructions on the package until it is malleable. The Polymorph should become transparent when it is ready to be sculpted.

a. If you used the water heating method, carefully remove the Polymorph from the hot water with tongs, and squeeze out the excess water. You can either sculpt the Polymorph by hand or use a form to get you started.

If you are using a form carefully shape the Polymorph around the form. While the Polymorph is still transparent, remove it from the form and sculpt it around the LED. Set it aside to let it harden.

If you are not using a form carefully shape the Polymorph directly around the LED until you get a desired shape. Set aside to let it harden. Repeat for other LEDs, creating a variety of shapes and designs.

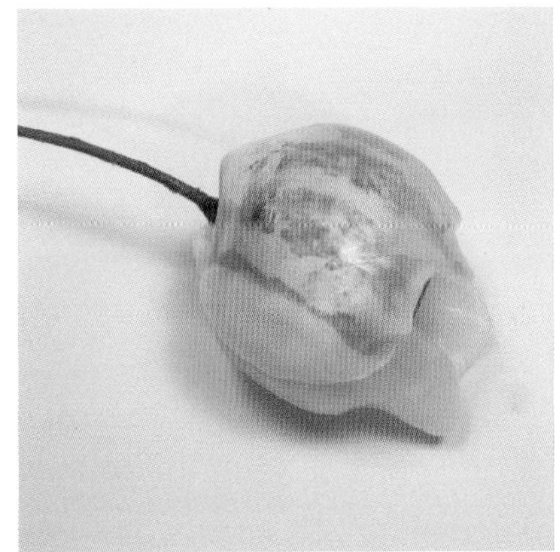

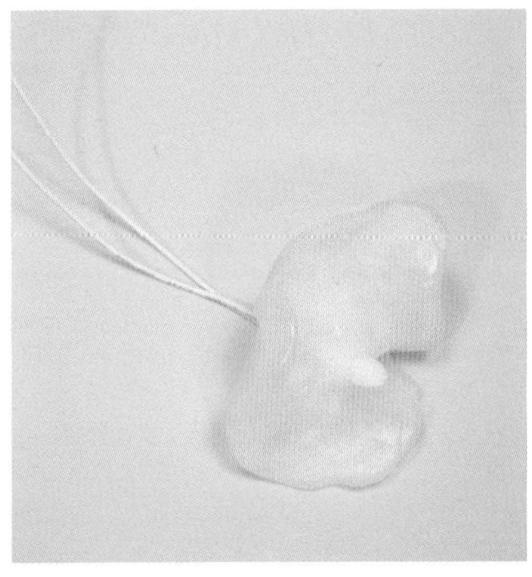

TIP: Polymorph plastic, just like clay, can be kneaded flat and then hand-sculpted into a three-dimensional form. Begin by flattening a small amount of Polymorph into a circle 2" in diameter. Center the LED in the middle of the Polymorph and pinch the Polymorph around the LED. Using your thumb and forefinger, shape the Polymorph until you are satisfied with the form.

b. Now cut all sets of LED wires to different lengths, with a minimum length of 12". Using wire cutters and strippers, for each LED, first cut and strip the negative wire and, using a black permanent marker, mark the end of the wire. Repeat, cutting and stripping the positive wire to the exact same length as the negative. Repeat for all LEDs, cutting the wires at varying lengths.

c. For each LED, twist the 2 wires loosely together. Add a touch of Polymorph along the length of the twisted wires to hold them in place. Repeat for all LEDs.

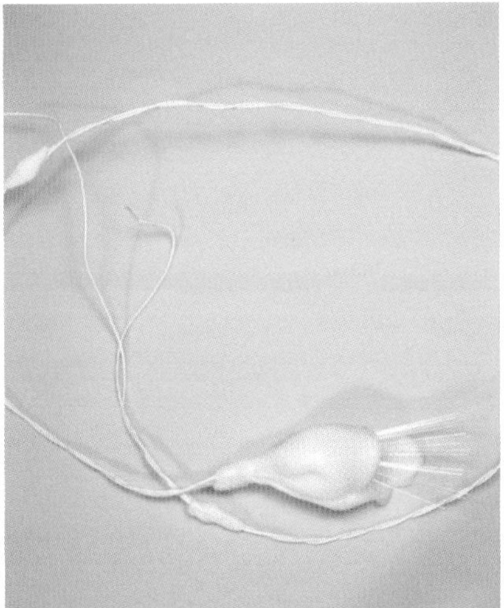

6. MAKE THE HANGING FRAME

a. Using a pencil and a fabric tape ruler, mark 12 drilling guides approximately 2.35" (2 7/20") apart around the circumference of the embroidery hoop.

b. Clamp the embroidery hoop onto a piece of wood. Using a hand drill and the 3/16" drill bit, drill holes around the hoop on the designated marked lines.

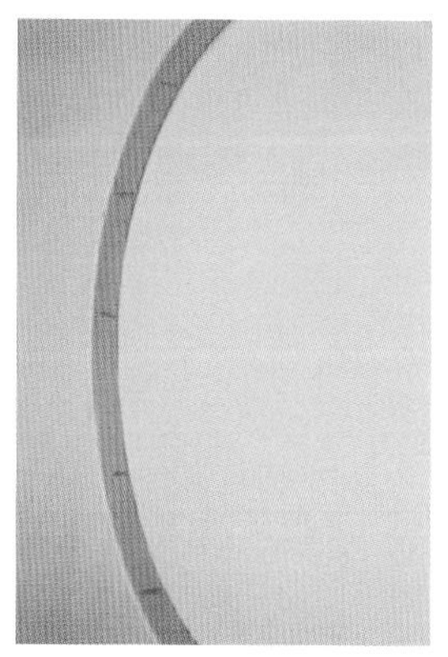

c. Slip the dowels through the drilled holes, placing them in the center of the loop, perpendicular to each other. They will fit snugly in the holes. Then paint the frame using white brush-on paint or spray paint. Set aside to let it dry.

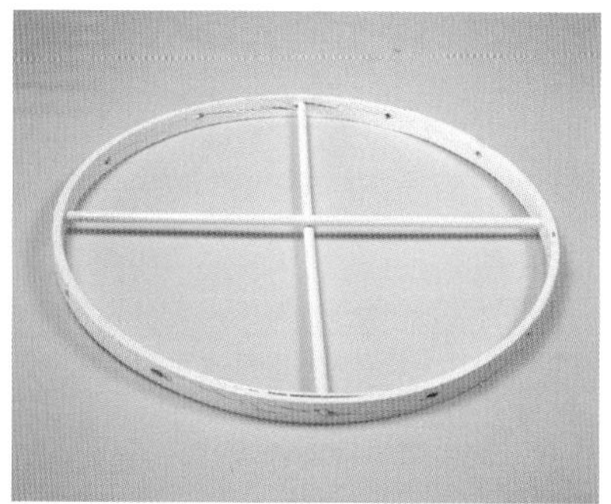

7. STRING THE LEDS TO THE HANGING FRAME

The LED Chandelier has a cluster of LEDs hanging from the middle of the frame as well as around the frame. The LEDs positioned around the frame must be balanced in order for the chandelier to hang properly. Before you start stringing the LEDs into the frame, you need to first determine which LEDs you want placed in the center and which should go around the circumference of the frame.

a. Take an LED that you would like to position on the frame's circumference. String the LED wires, starting from the inside of the frame through the drilled holes and out over the top edge. Pull the wires towards the center, knotting them around the cross section of the dowels. Make sure to leave at least 3" of wire at the end of the knot. Repeat, stringing an even set of LEDs on opposite sides of the frame.

TIP: The best placement consists of hanging the longer LEDs in the center, clustered with the shortest, placing the medium-length LEDs around the circumference.

b. Next, take one of the LEDs that you would like to position in the center of the frame. Knot the wire around the cross section of the dowels, making sure to leave at least 3" of wire at the end of the knot. Repeat, stringing the remaining LEDs to the center of the frame.

c. Separate the 16 negative wires from the 16 positive wires. You can temporarily hold them in place with a rubber band. Then separate the 16 negative wires into 5 separate clusters (one cluster will have 4 wires). Using wire cutters, cut the wires of each cluster so that they are approximately the same length. Using wire strippers, strip the wire ends. Twist the ends of each wire cluster together. Cut a piece of 3/16" heat-shrink tubing and slip it over each cluster, making sure that the stripped ends remain exposed. Solder the stripped ends of each wire cluster together.

d. Using wire cutters, cut 5 pieces of 22 AWG stranded wire 36" in length. Using wire strippers, strip about 1" off each end. Wrap one of the stripped ends of a wire around one of the soldered wire clusters. Solder. Repeat, soldering the remaining wires to each cluster.

e. Slip the heat-shrink tubing over the exposed metal of each cluster. Using a heat gun, shrink the tubing until it is snug. Secure the wires temporarily together using a rubber band. You should now have 5 negative wires, reduced from the previous 16. Repeat the process for the positive wires, separating them into clusters and soldering a 36" long piece of stranded wire to each cluster.

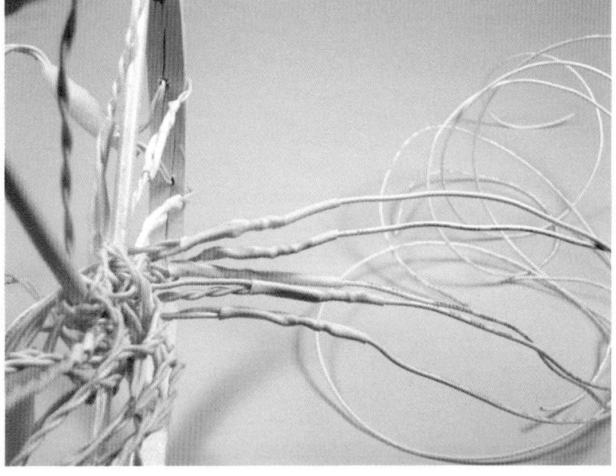

NOTE: Make sure to note which are the positive and negative wires so you don't confuse the two. Mark all the negative wires with a black marker.

f. Cut a 2" piece of the 3/8" heat-shrink. Slip it over the 5 positive wires. Twist the stranded ends of all 5 wires together.

g. To create the magnetic switch, you will first connect one of the magnet switches to the 5 positive LED wires and then the second magnet switch to the positive power adapter/lamp cord wire. When the 2 magnets connect, the circuit will be closed and the electricity will flow, lighting the LEDs. Take one of the magnet switches and twist the stripped end of the magnet wire to the stripped ends of the 5 positive wires. Solder. Slip heat-shrink over it and, using a heat gun, shrink the tubing. Set aside the second magnet switch. You will connect it later to the positive wire of the power adapter.

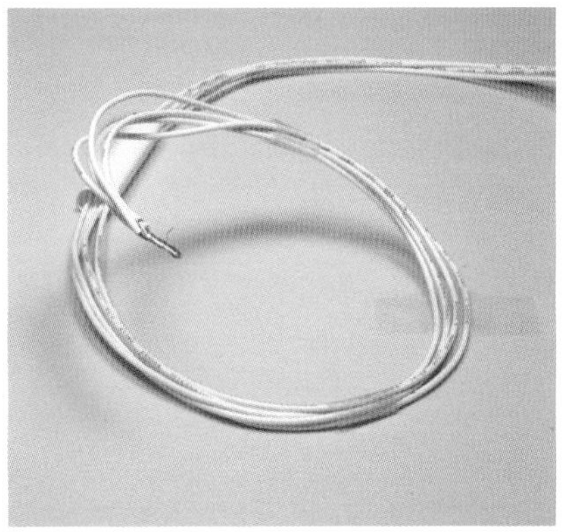

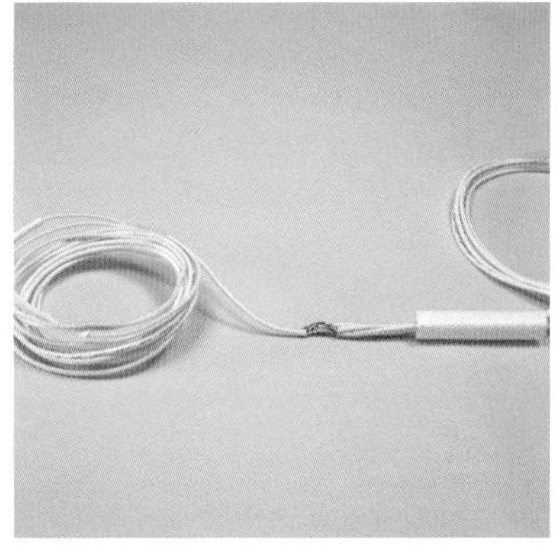

8. WIRE THE POWER ADAPTER

a. First determine the polarity of the power adapter wires by following the instructions on page 43. Once you have determined the polarity of the wires, make sure to mark the negative wire with either a piece of electrical tape or with a colored marker. Separate the 2 power adapter wires about 3". You will be extending the power adapter wires with the addition of the white lamp cord.

Using wire cutters, snip the center of the white lamp cord and pull the wires apart about 3", separating the 2 wires. Using wire strippers, remove about ½" of insulation. Repeat for the opposite side. Designate which wire you will be using as the ground wire and mark both ends with a black marker.

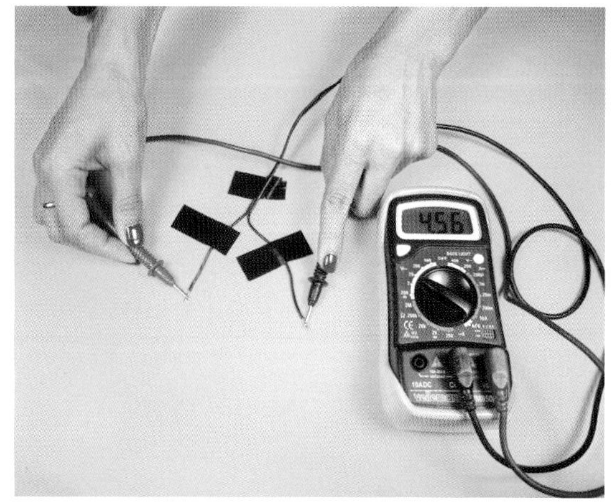

b. Cut two 1" pieces of the 3⁄16" heat-shrink tubing. Slip the heat-shrink over each wire extending from the power adapter. Twist the negative ground wire of the power adapter to the negative ground wire (marked with black) of the white lamp cord. Repeat for the positive wire. Solder the wires. Using a heat gun, shrink the tubing.

c. Cut a 6" piece of the 3⁄8" heat-shrink tubing. Slip the heat-shrink from the opposite end of the lamp cord wire over the 2 joined wires and shrink the tubing.

d. Now that you have connected the power adapter to the lamp cord, the next step is to connect the positive lamp cord wire (on the opposite end) to the remaining magnet switch. Cut a 4" piece of the 3⁄8" heat-shrink tubing. Slip the heat-shrink tubing over the lamp cord wires.

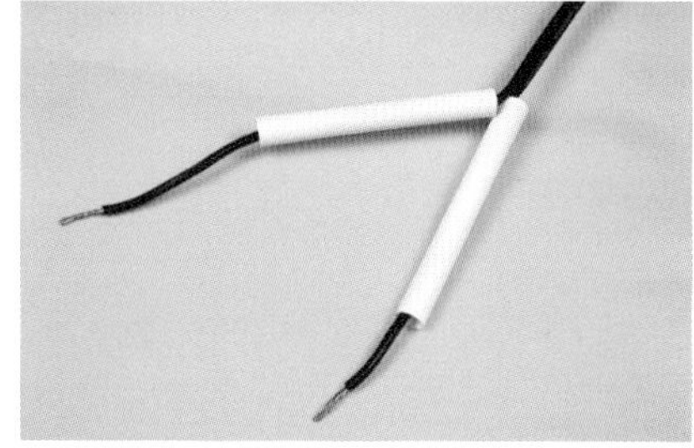

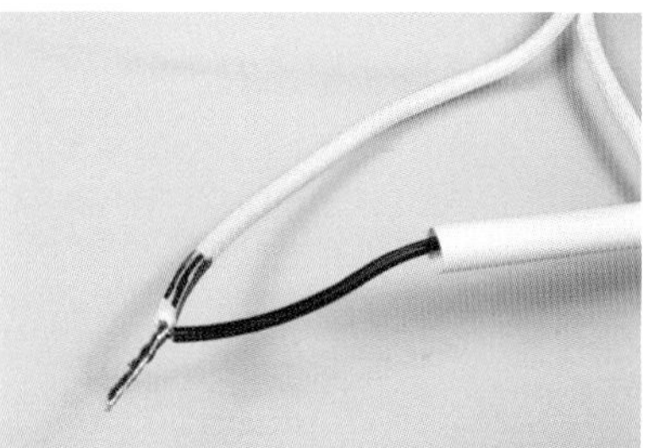

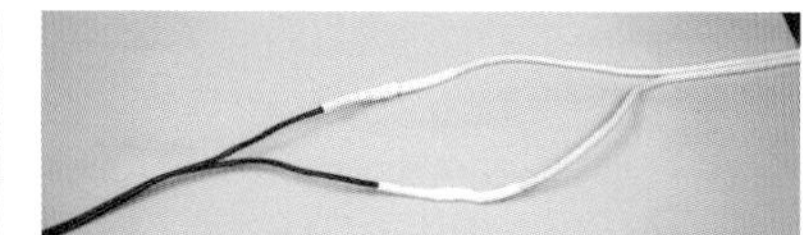

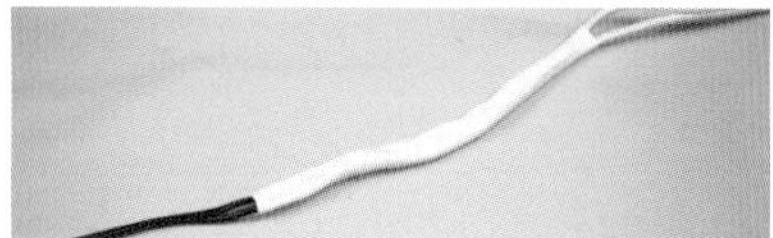

e. Cut a 1" piece of 3⁄16" heat-shrink tubing and slip the tubing over the magnet wire. Twist the stripped end of the remaining magnet switch to the positive wire and solder. Slip the 3⁄16" heat-shrink over the exposed metal and shrink until snug.

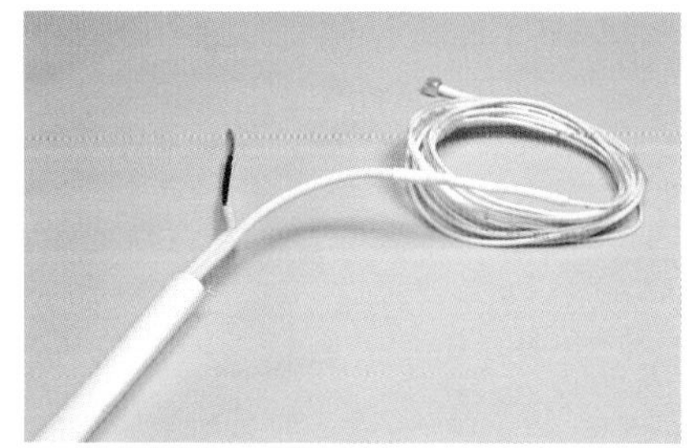

f. Next you need to connect the negative LED wires to the negative lamp cord wire. Grab the cluster of negative LED wires. Cut a 1" piece of 3⁄16" heat-shrink tubing and slip the tubing over the negative lamp cord wire. Twist the stripped ends of the LED wires, connecting them to the negative lamp cord wire. Solder. Slip the heat-shrink over the exposed connection and shrink the tubing. Slip the 4" piece of 3⁄8" heat-shrink over all the connected wires and shrink the tubing.

g. Using the zip tie, fasten the negative LED wires (now connected to the lamp cord) and the positive LED wires (now connected to a magnet switch) together. Gently twist all wires, centering the wire cluster to give the chandelier proper alignment.

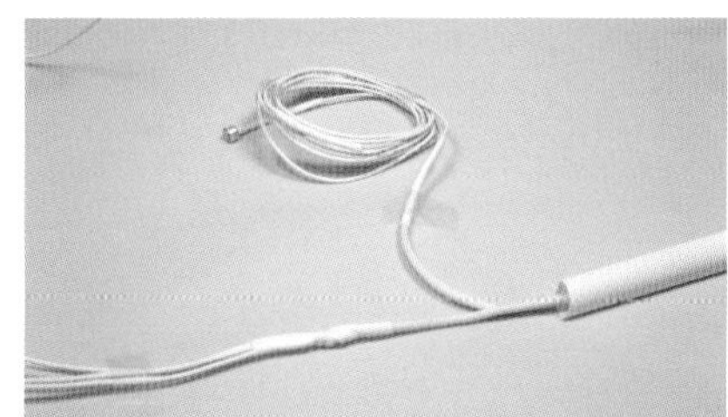

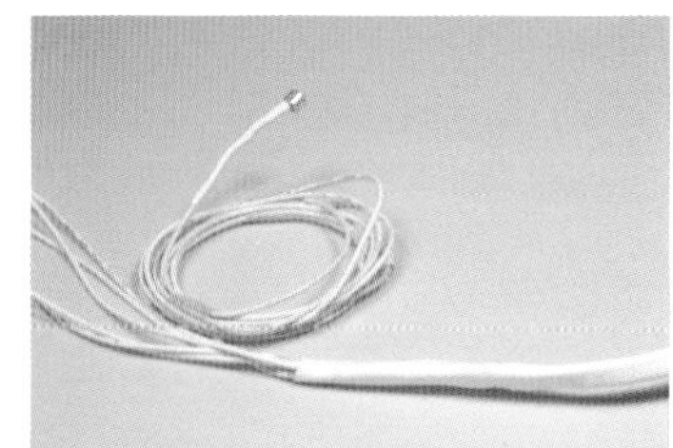

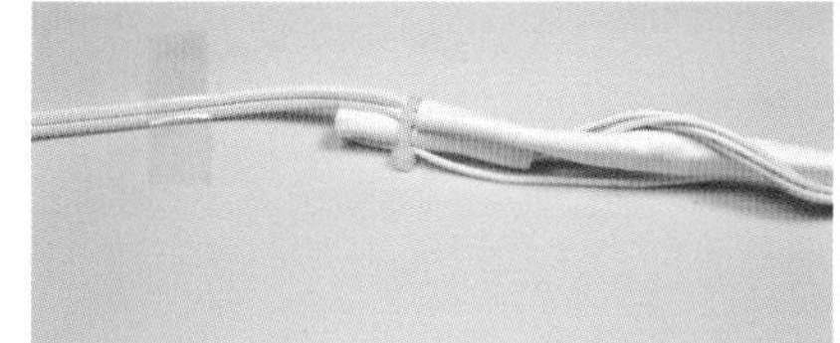

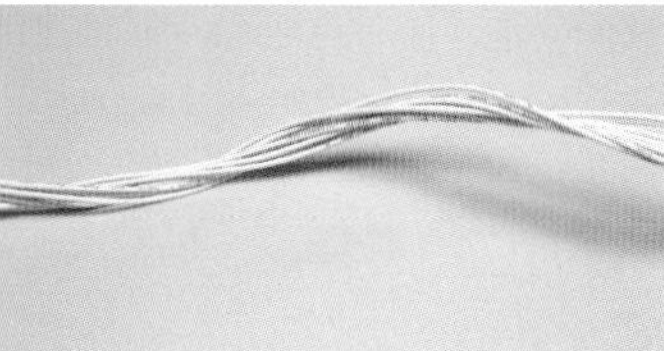

9. MAKE THE SHADE

a. Cut the decorative fabric or paper into a 30"×7¼" piece. Place the fabric wrong side up on a table.

b. Cut the 60" of boning into two 30" pieces. Using the appropriate glue for the material, glue 1 piece of boning to the top edge of the fabric. Glue the second piece ½" above the bottom edge of the fabric. Let dry.

TIP: Use clothespins or small clips to hold the boning in place while the glue dries.

c. Using a ruler, place marks 1½" apart on the bottom edge of the fabric.

d. String the beads, each with a jump hoop. Loop the beads through the fabric around the designated marks.

e. Place a bead of glue on the outside of the hanging frame. Carefully glue the beaded fabric around the circumference of the frame. The beginning and end of the fabric should overlap by about ½". Now add a touch of glue along the overlapping fabric (vertical) seam. Let dry.

FINISH

ROCK 'N' ROLL SPEAKERS

Whether you're in a hostel in Bangkok or backpacking in Goa, you can rock out to your favorite tunes with the portable Rock 'n' Roll Speakers. Powered by a 9V battery, the Rock 'n' Roll Speakers are a perfect, compact travel companion for adventurous jet setters.

HOW IT WORKS »

The Rock 'n' Roll Speakers are a set of customizable, lo-fi travel speakers that allow you to take your music with you everywhere. This project consists of a power amplifier circuit and a tilt switch. With a gentle rock, the portable speakers are powered on and off. When both speakers are visible, the speakers are "on," and when only one speaker is visible, the speakers are "off."

COMPLEXITY: ADVANCED

WHAT YOU WILL LEARN

- Build a circuit using a perfboard
- Work with ICs

RELATED TUTORIALS

- Art of Soldering (45-49)
- Building a Simple Circuit (27-33)
- Prototyping a Circuit Using a Perforated Board (33)

MATERIALS

- 18"×18" mat board
- Double-sided tape
- Hook and loop square, 1"
- Hook and loop strip, 51"
- Glue
- 18"×18" decorative paper or fabric for speaker cover (optional)
- White spray paint (optional)

TOOLS

- Scissors
- Utility knife
- Needlenose pliers
- Wire cutters
- Wire strippers
- Hot glue gun
- Heat gun or blow dryer
- Marker

ELECTRONICS

- 220µF capacitor
- 0.05µF capacitor
- 10Ω resistor
- LM386 audio amplifier IC
- 10KΩ trimpot variable resistor
- 8Ω speakers (2)
- 1/8" audio plug
- Tilt switch nonmetallic
- 9V battery connector
- 9V battery
- 2"×2" perfboard
- 6' stranded wire
- Variety of jumper wires or solid wire
- 1/6" heat-shrink tubing, 8"
- 1/4" clear heat-shrink tubing, 16"

TEMPLATES

Template available for download at fashioningtechnology/speakers.

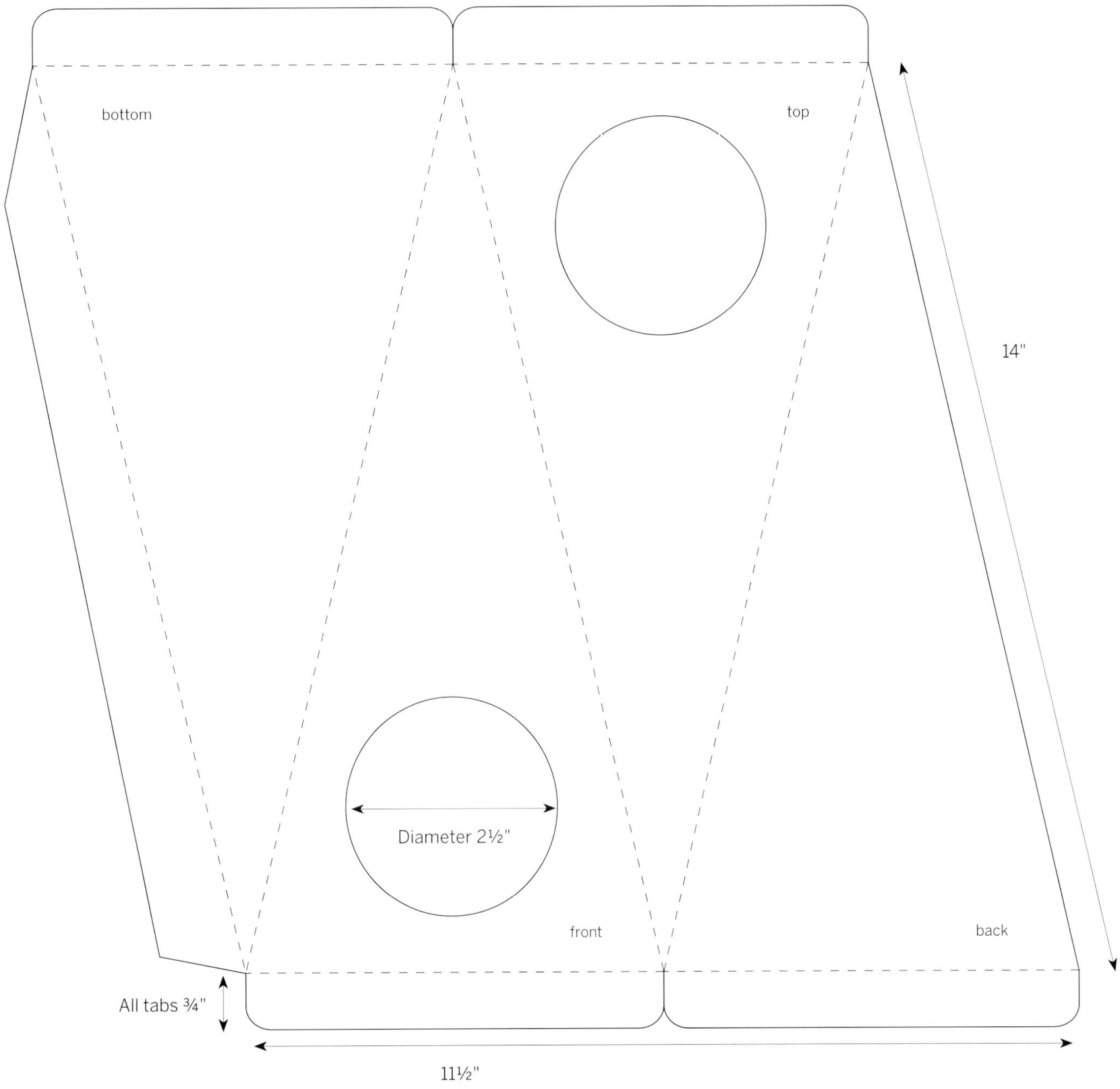

START »

1. MAKE THE FORM

a. Using the provided template, trace the shape onto mat board and cut.

b. You can optionally customize the speakers by covering the form with decorative paper or fabric. Using the appropriate adhesive for the selected material, glue it directly on top of the cut form, wrapping the edges into the inside of the form.

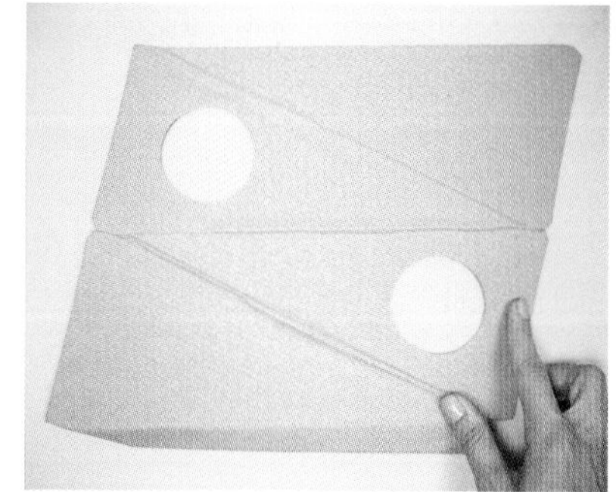

c. Fold the form along the score lines. If you choose not to cover the form with paper or fabric, you can also paint it. Then, using spray paint, spray the speakers until they are evenly coated. Painting the speakers is optional.

2. MAKE THE SWITCH

a. Cut and strip both ends of two 1½" pieces of stranded wire. Next, cut two 1" pieces of ⅙" heat-shrink tubing.

b. Grab the tilt switch. Wrap one of the stripped ends of a wire on a lead and solder. Slip the heat-shrink tubing over the connection. Using a heat gun or blow dryer, shrink the tubing. Repeat for the second lead.

c. Cut the positive, red wire of the connector in two, ½" from the top. Strip the cut ends.

d. Next, cut two 1" pieces of ⅙" heat-shrink tubing. Slip the pieces of heat-shrink over one of the switch wires. Twist the 2 wires from the switch and battery connector together and solder. Slip the tubing over the connection and shrink. Repeat, connecting the second switch wire to the loose, cut wire from the battery connector.

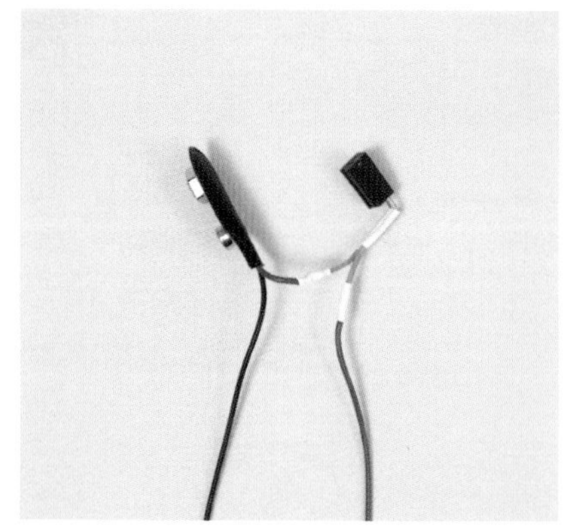

3. MAKE THE PLUG

a. Cut and strip both ends of three 16" pieces of stranded wire. Twist the ends.

b. Remove the audio plug cover. Slip the wires through right, left, and center terminals of the plug and twist them into place. Solder.

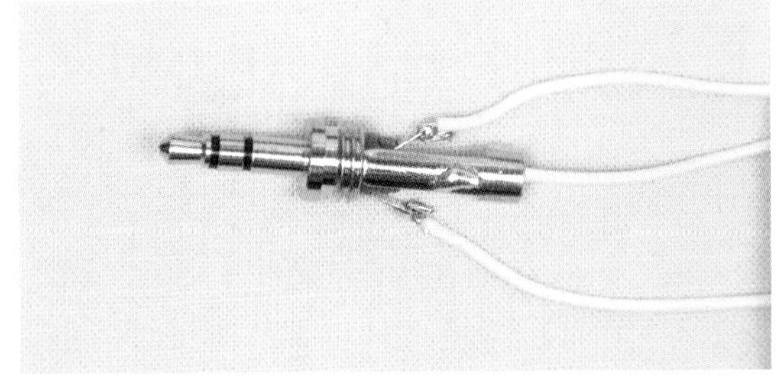

c. Cut two 1" pieces of ⅙" heat-shrink tubing. Slip them over the right and left terminals and shrink.

d. Using a black marker, mark the bottom of the wire leading from the center terminal. This will be the ground wire. Slip the ¼" clear heat-shrink tubing over all the wires. DO NOT SHRINK. Replace the plug cover.

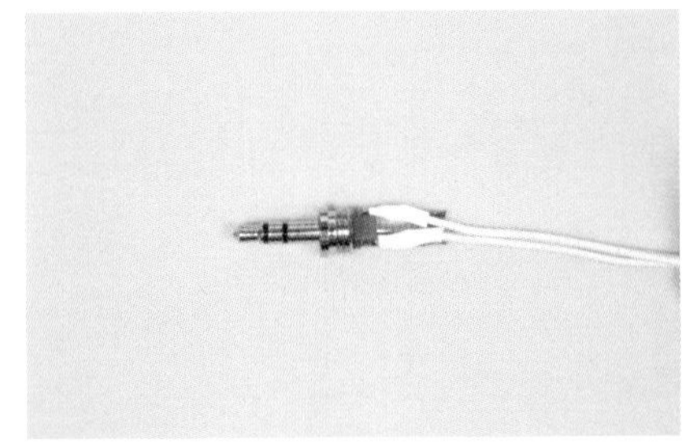

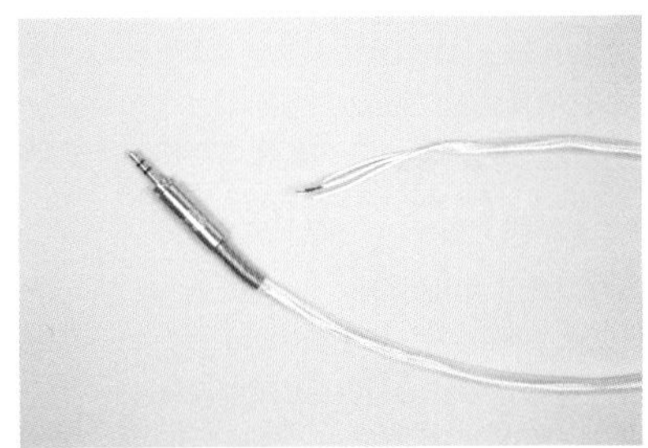

4. BUILD THE CIRCUIT

If you have never built a circuit on a perforated board before, review page 33 for more detailed information on how to use a perfboard. The illustration at right shows how the components of the circuit are connected to each other. It is not a direct translation on how the circuit will be played out on the perfboard.

a. Grab the LM386 audio amplifier IC. Locate the notch and/or dot. The pins on the IC are numbered starting counterclockwise from the dot, as in the illustration. Using a perfboard designed with standard IC and component spacing, place the IC in the center of the perfboard. Designate a row for power and ground by placing a "+" and "–" near the assigned holes.

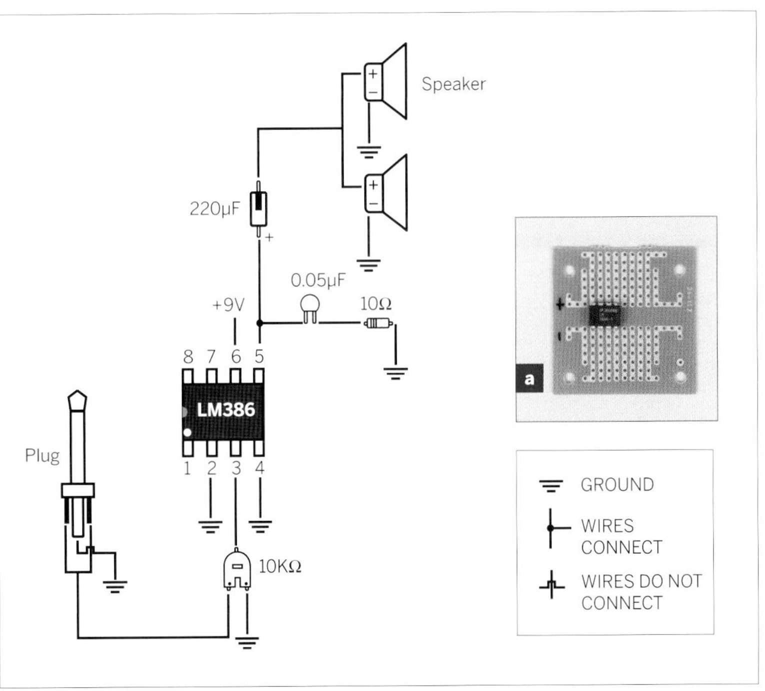

b. Using jumper wire, connect pin 6 to power, pin 4 to ground, and pin 2 to ground.

c. From pin 5, a 0.05μF ceramic capacitor must connect to the 10Ω resistor, which then has to be grounded. Also from pin 5, a 220μF capacitor must be connected. Start by connecting a wire from pin 5 and jump it 2 rows over (red wire) to the right. Connect 1 lead of the 0.05μF ceramic capacitor to the row the red wire has been jumped to and the other lead 2 rows over to the right. To hold the capacitor temporarily in place, bend the leads flush to the board. Connect 1 lead of the 10Ω resistor to the second lead of the 0.05μF ceramic capacitor and the second lead of the 10Ω resistor to a separate empty row. Using jumper wires, connect the second lead of the resistor to ground.

d. Connect another wire from pin 5 and jump it 3 rows over (yellow wire). Connect the positive lead of the 220μF electrolytic capacitor (the one without the black band) to the row the yellow wire has been jumped to. Connect the negative lead of the 220μF electrolytic capacitor 2 rows over to the right.

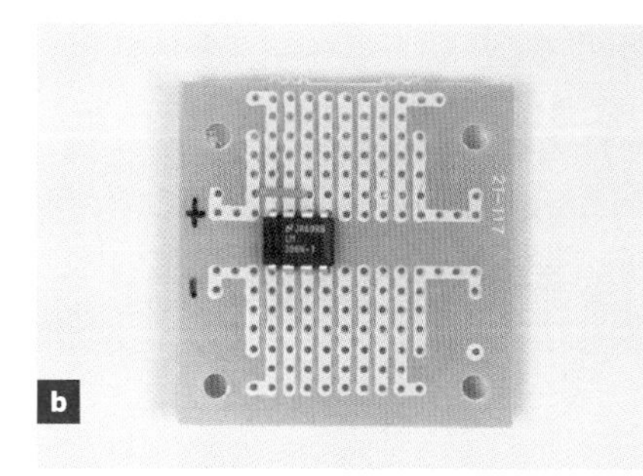

b

b

b

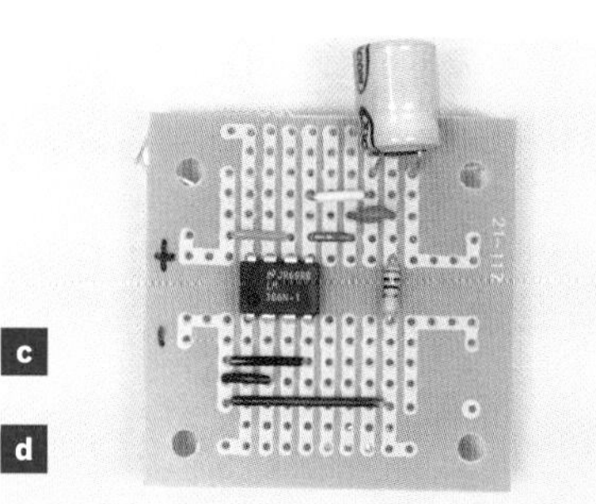

c

d

e. Ultimately, the middle lead of the 10KΩ trimpot must be connected to pin 3 with one of its outer leads connected to ground and the other connected to the input of the audio plug. Start by connecting a wire from pin 3 and jump it 3 rows over (blue wire). Connect the middle lead of the trimpot to the row the blue wire has been jumped to and the 2 other leads to empty rows. Connect the right lead of the trimpot to ground. The small black jumper wire (to the right of the trimpot) is connected to a row that is grounded.

f. Cut and strip six 8" pieces of stranded wire. Take 2 wires and connect them to the negative lead of the 220μF electrolytic capacitor. The wires will be the positive wires of the speakers.

e

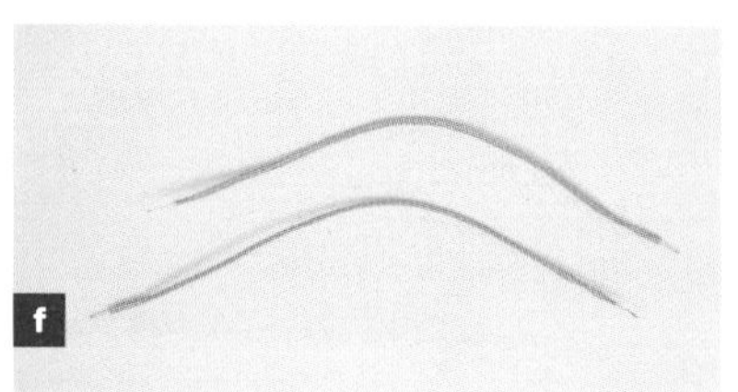

f

f

g. Grab another 2 wires. Connect the wires to ground. Using a black marker, mark the 2 wires. The wires will be the negative wires for the speaker.

h. Next, connect a wire to the second (not grounded) lead of the trimpot. This wire will lead to both plug inputs. Grab the last wire. Connect it to ground. This wire will lead to the plug ground (marked with black).

i. Take the battery connector-switch piece, and insert the red wire into the power row and the negative into ground. Now grab the speakers. Slip the positive speaker wires from the board (Step 4f) into the positive terminal of the speakers and the negative wires (Step 4g) into the negative terminals. Twist, securing them into place.

j. Carefully review all the connection points in the circuit. Once you are certain that everything is connected properly, solder all the components into place. Make sure to solder the connections at the speaker terminals as well.

NOTE: The audio plug should not be soldered to its respective positive and negative wires on the board. You will add the audio plug at a later time.

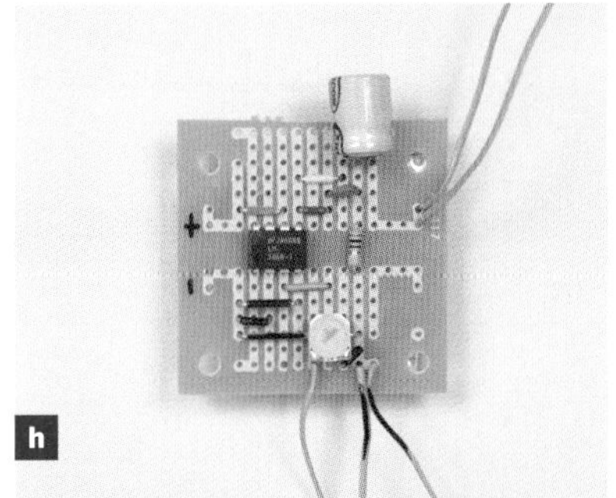

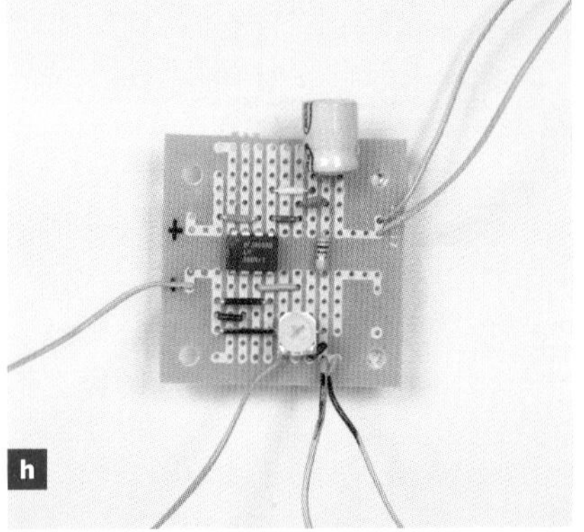

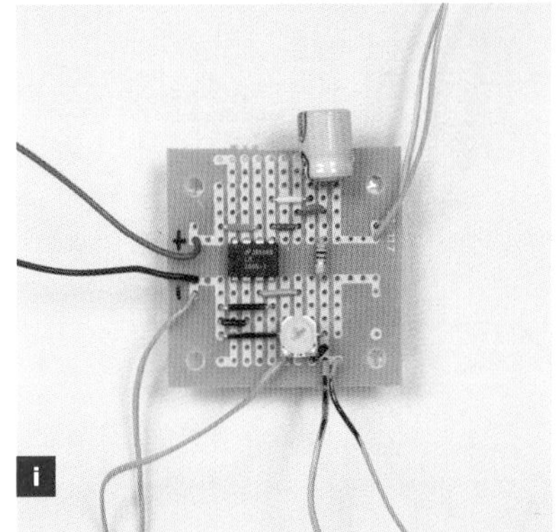

5. ATTACH THE SPEAKERS AND CIRCUIT TO THE FORM

a. Grab the speakers and circuit. Using hot glue, attached the speakers in place on the form. Position the circuit on the front triangle and glue it securely to the mat board.

b. Cut two 1" pieces of 1⁄6" heat-shrink. Slip the heat-shrink over the ground wire of the plug. Wrap the ground wire of the audio plug to the ground wire coming from the circuit (Step 4i). Solder and shrink the tubing. Repeat for the other 2 input wires, attaching both input wires from the plug to the remaining wire connected to the second lead of the trimpot.

c. Locate an ideal position for the 9V battery on the bottom triangle of the form. Using a piece of sticky hook and loop, attach the battery to the bottom triangle. Using hot glue, adhere corresponding strips of hook and loop on opposite ends of the inside of the form and the top of the fold.

d. Connect a 9V battery to the battery connector and plug in your portable music player.

e. Determine the angle that the tilt switch is "on." Hot glue the tilt switch in the appropriate "on" position when the front of the speaker is facing forward (when you can see both speakers) and "off" when the front speaker is facing down (when you can only see 1 speaker).

f. Close up the form, having the plug extend out from the back corner. Now you're ready to rock 'n' roll wherever you go!

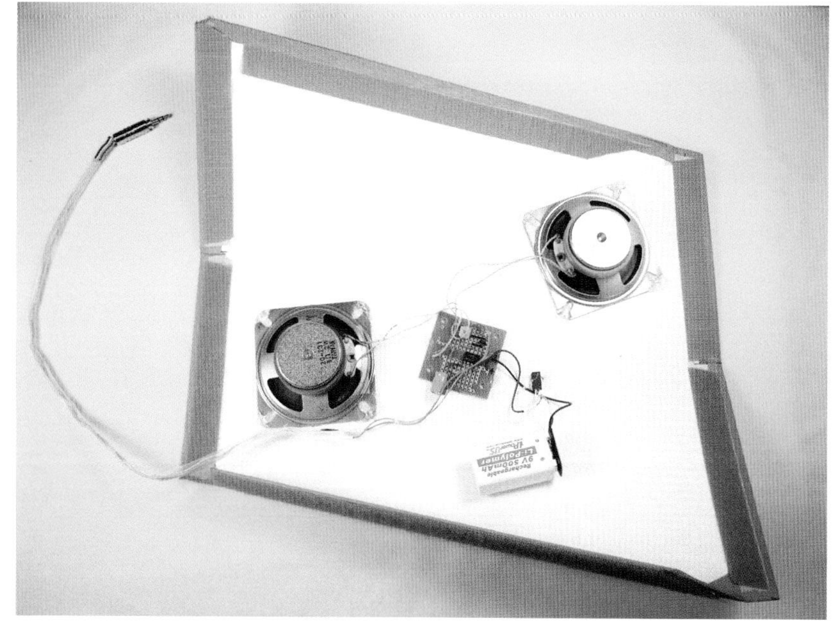

FINISH

Now Playing
2 of 14
Jump Up Jump Up
JT And The Clouds
The City's Hot Yeah...
MENU

5

// **Toys aren't just for tots; they're for adults, too. Embrace your inner child by crafting interactive toys imbued with unique, whimsical personalities.**

INTERACTIVE TOYS

With the addition of simple circuitry, you can create a handful of interactive finger puppets that pulse with emotion. Or craft a biotope of foldable, paper glow bugs that luminesce in the dark. From enchanting mobiles that periodically twitch to solar pull toys that translate the gleaming rays of the sun into delightful song, the toy tutorials in this chapter are designed to cultivate your imagination and inspire creative play.

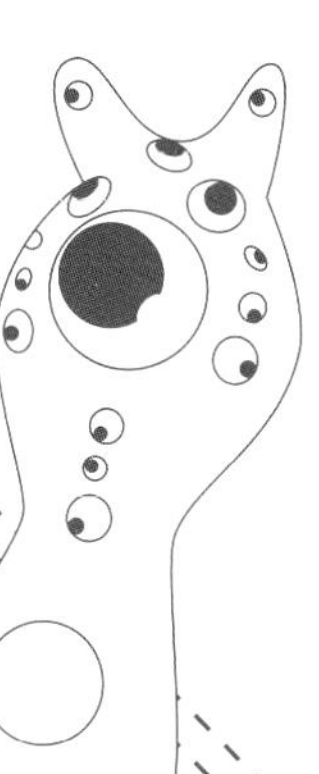

E-PUPPETS

EMOTING ELECTRONIC FINGER PUPPETS

Meet Tremble, Blush, and Spitfire — the darlings of interactive finger puppet theater. Emotional and expressive, each E-puppet has his own personality quirk. Tremble is a little neurotic and fearful, always watching his back. Blush, the cutest, is timid and easily embarrassed. And the alpha, Spitfire, is the protector, plagued with a hot temper. The three together provide hours of comedic entertainment and imaginative play.

HOW IT WORKS »

With the addition of a few electronic components — a vibrating motor, piezo speaker, or a couple of LEDs — you can embellish an ordinary finger puppet and give it the extraordinary ability to express the subtlest of emotions. Each E-puppet contains a middle felt layer sewn with a different simple circuit. The circuit is activated by pressing on the puppet's belly button. The puppets are all powered by the same battery source located on the bracelet.

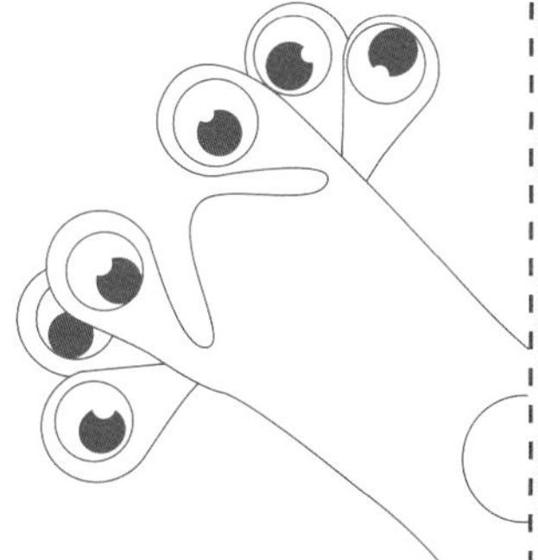

Once you're familiar with how your puppets are constructed, why not grow the family? Using simple circuits, you can create a new puppet that can squawk in protest (piezo speaker), spin his tail in excitement (motor), or have his heart pulse with nervousness (blinking LED). The opportunities are endless!

COMPLEXITY: BASIC

WHAT YOU WILL LEARN

- Sew simple soft circuits
- Make soft switches

RELATED TUTORIALS

- Working with LEDs (19-23)
- Sewing Soft Circuits (57-67)

MATERIALS

- **Colored wool or wool felt** 3 colors, 1/3yd each
- **Conductive thread**
- **Clear/white fiber optics, 4'**
- **Rubber band**
- **Regular thread**
- **Large button**
- **Stranded wire, 5'**
- **Solder**
- **Metal snaps (9 pairs)** female and male
- **Metal loops (4)** eye part of hook and eye jewelry closure
- **Conductive fabric or conductive fabric tape, 3"**
- **Colored poofs (3)** available at craft stores
- **Glue**
- **Googly eyes** of various sizes
- **Embroidery floss**
- **Stuffing**

TOOLS

- Soldering iron
- Scissors
- Sewing needle
- Needlenose pliers
- Wire cutters
- Wire strippers
- Pencil or marking pen
- Computer and printer
- Hot glue gun
- Sewing pins

ELECTRONICS

- **Vibrating motor** for Tremble, part # VPM2, solarbotics.com
- **3mm red LEDs (2)** for Blush
- **5mm red LED** for Spitfire
- **3V lithium photo battery**
- **3V battery holder**

TEMPLATES

- Template A
  Top, back, and middle circuit layer for Tremble (3)
- Template B
  Top, back, and middle circuit layer for Spitfire (3)
- Template C
  Top, back, and middle circuit layer for Blush (3)
- Template D
  Face layer for Blush (1)
- Template E
  Power source bracelet layers (2)

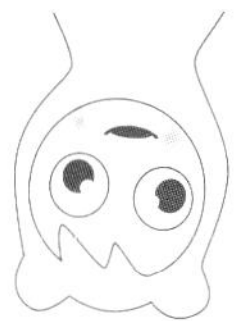

TEMPLATES

All templates, as well as sewing guides for Spitfire and Blush, can be found online at fashioningtechnology.com/epuppets.

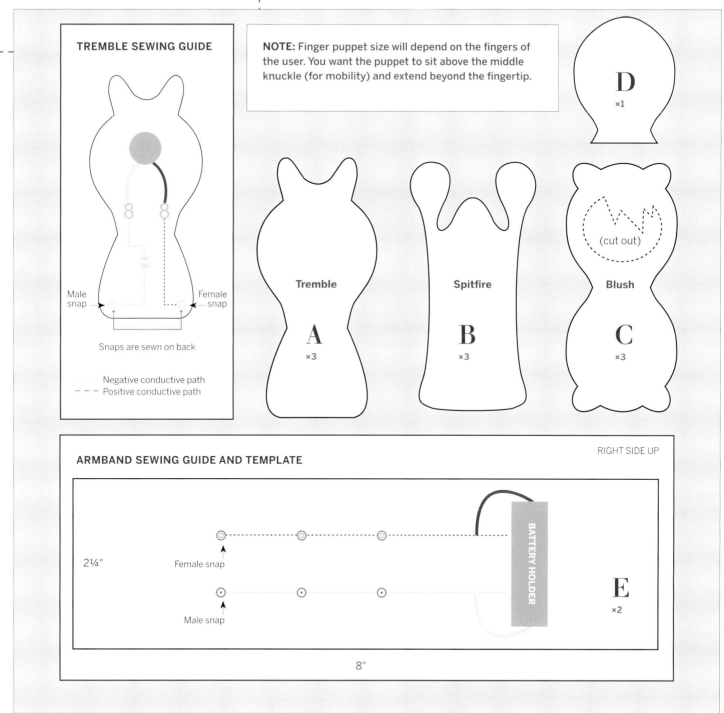

TREMBLE SEWING GUIDE

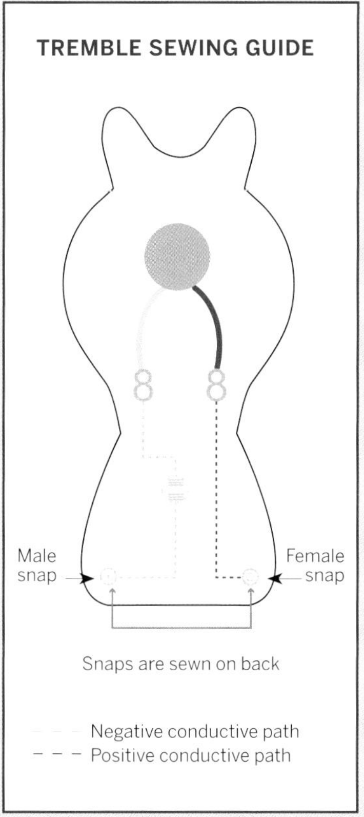

NOTE: Finger puppet size will depend on the fingers of the user. You want the puppet to sit above the middle knuckle (for mobility) and extend beyond the fingertip.

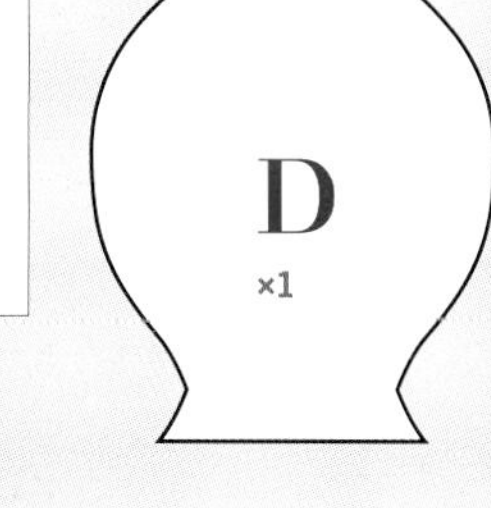

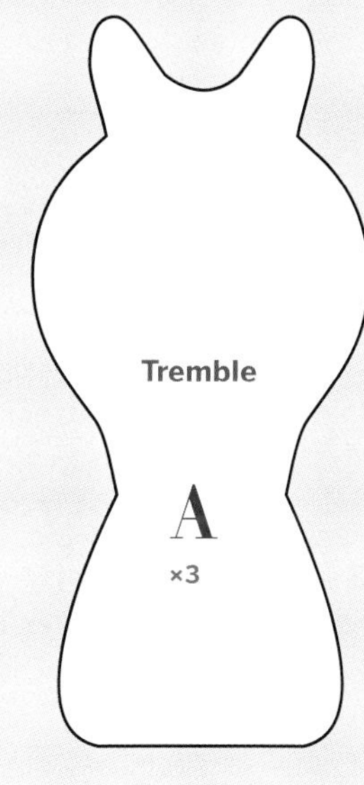

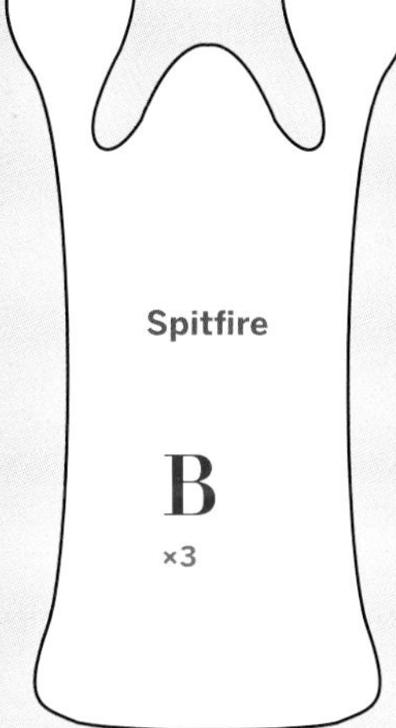

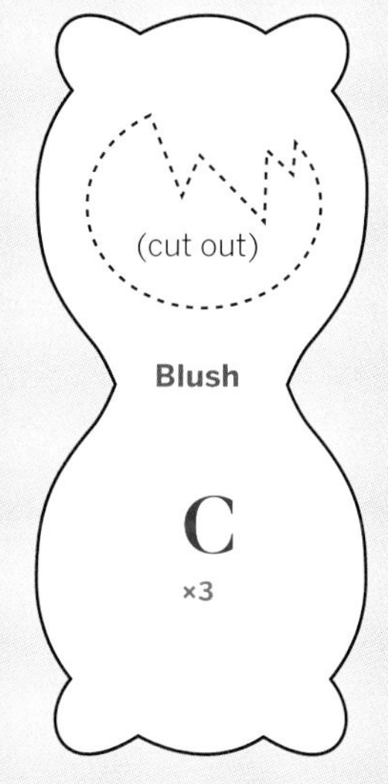

ARMBAND SEWING GUIDE AND TEMPLATE

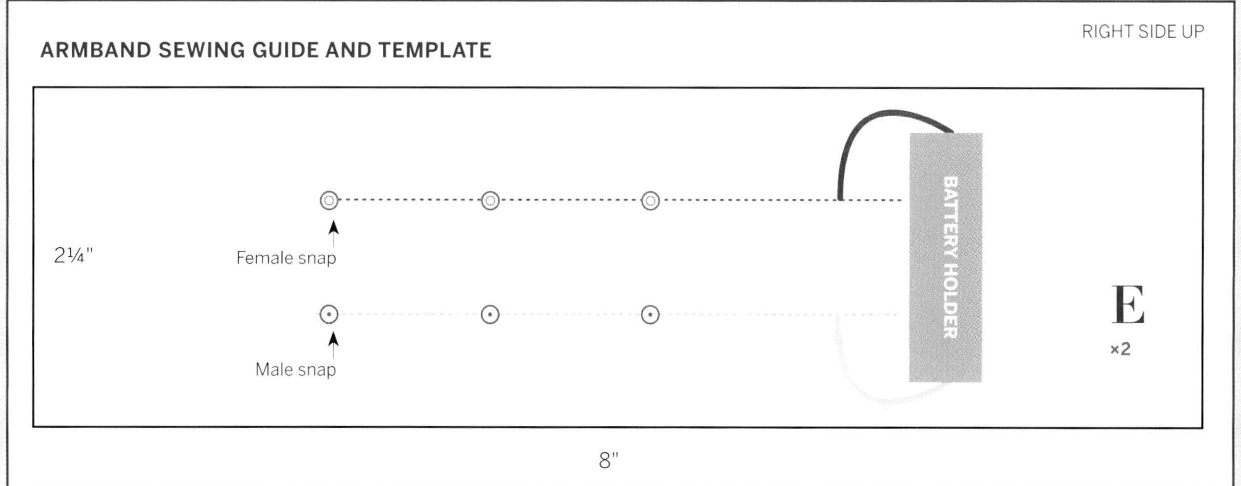

START » TREMBLE: Puppet with Vibrating Motor

1. CUT THE PATTERN

a. Print out Template A. Using a marking pen, trace the pattern three times onto the felt. Using scissors, cut out all 3 pieces of the template: the top layer, the middle (circuit) layer, and the back layer.

b. Using a hole punch, cut out a hole in the center of the top felt layer about ¾" from the bottom of the felt. This is your center switch mark. Place the top felt layer over the middle, circuit layer, making sure to align the edges. Using a marking pen, mark the location of the hole on the middle layer. You will be creating 2 contact points for your soft switch at this location.

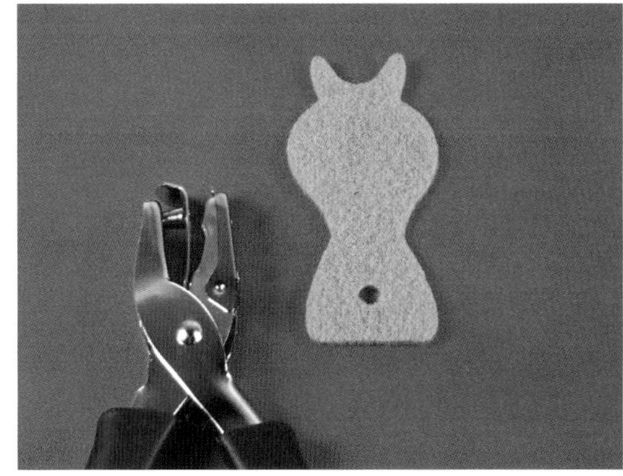

2. SEW THE CIRCUIT PATH

a. Take the vibrating motor and strip about ¼" of insulation off both ends of the wires. Loop the stripped end of the negative (blue) wire around the top circle of a metal jewelry eye. Using a soldering iron, solder the wires to the metal loops. Repeat for the positive wire.

b. Position the vibrating motor in the center of the upper portion of the middle felt layer. Using a sewing needle and conductive thread, loop the thread around the bottom loop of the negative vibrating motor wire. Continue to sew a straight line toward the center switch mark. Once you have reached the top of the mark, create the first contact point for the soft switch by embroidering a small patch of conductive thread (refer to the sewing guide on page 163). Knot and cut the thread.

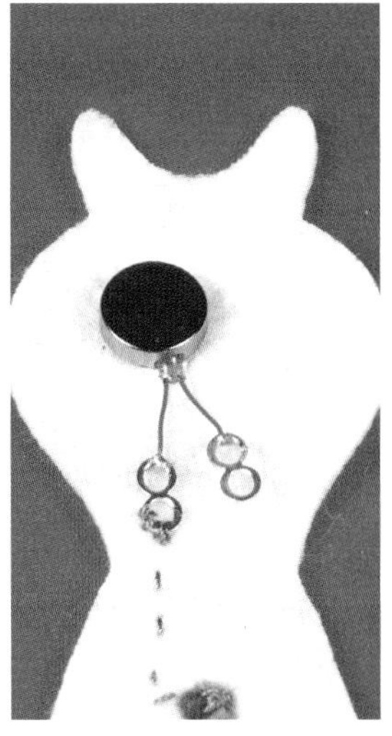

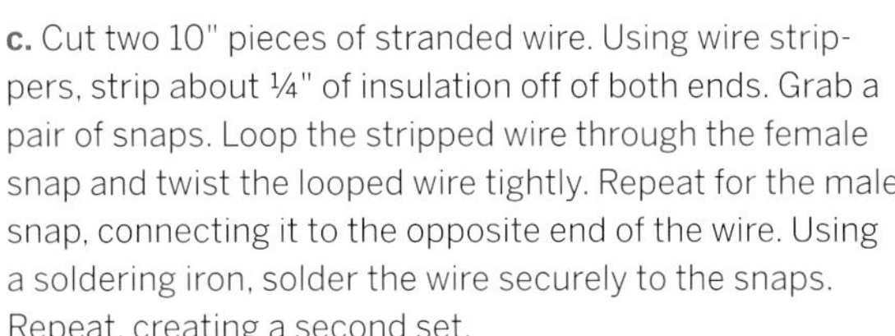

c. Cut two 10" pieces of stranded wire. Using wire strippers, strip about ¼" of insulation off of both ends. Grab a pair of snaps. Loop the stripped wire through the female snap and twist the looped wire tightly. Repeat for the male snap, connecting it to the opposite end of the wire. Using a soldering iron, solder the wire securely to the snaps. Repeat, creating a second set.

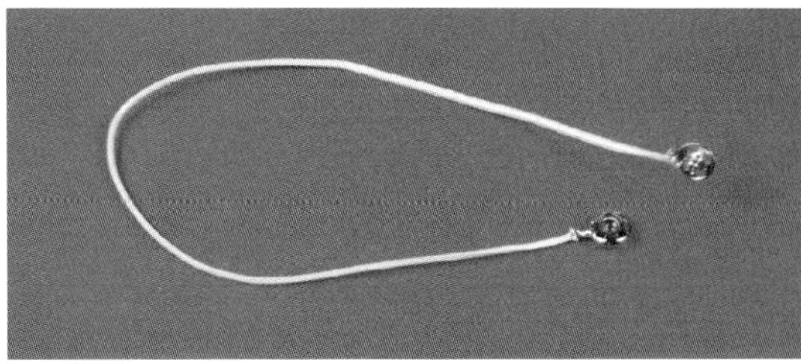

d. Using a sewing needle and conductive thread, create the second contact point for the soft switch by embroidering another small patch of conductive thread below the first. *These 2 patches of conductive thread should not touch.*

e. Continue to sew a path to the lower corner about ¼" from the bottom and side of the felt. Sew the male snap end of the wire securely onto the *back* of the felt (side without motor).

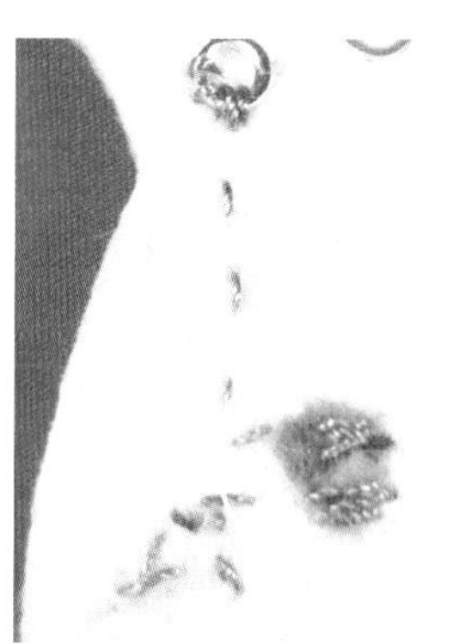

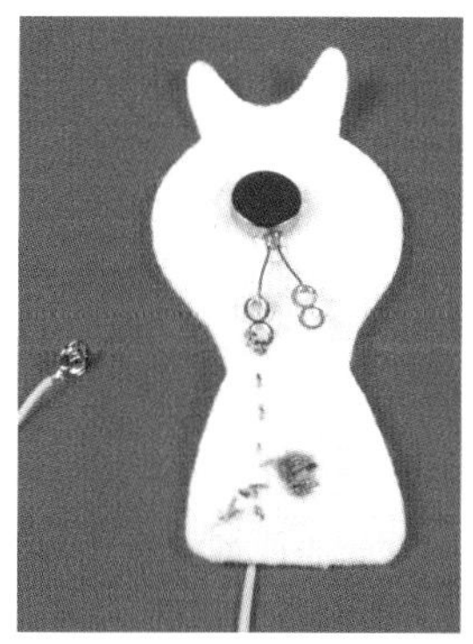

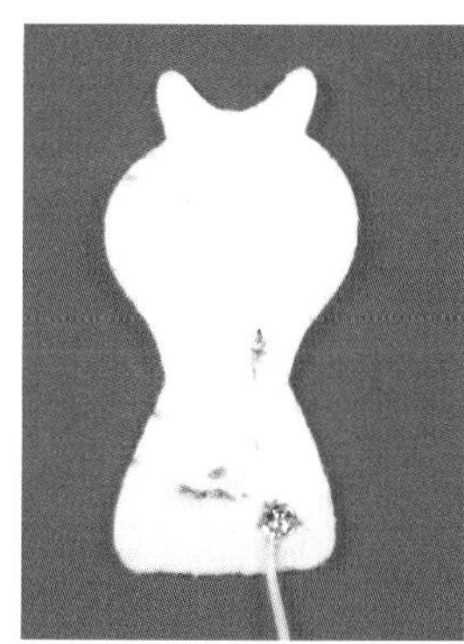

f. Using conductive thread, loop the thread around the bottom loop of the positive vibrating motor wire. Continue to sew a straight line toward the lower corner about ¼" from the bottom and side of the felt. Sew the female snap end of the second wire securely on the *back* of the felt.

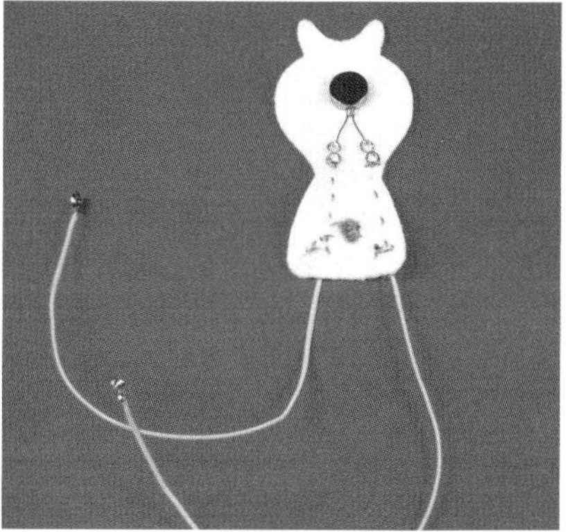

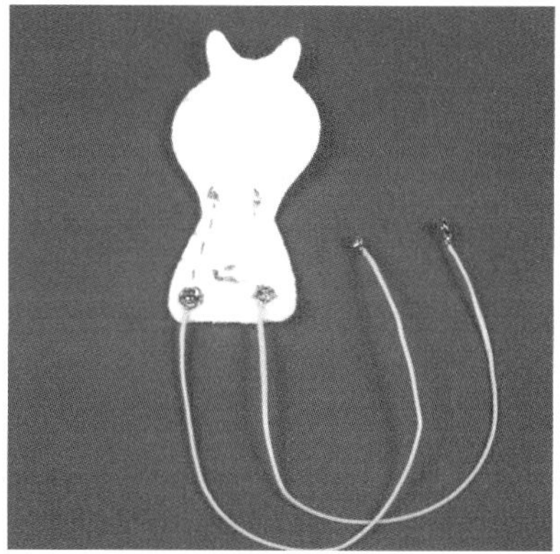

3. CREATE THE SOFT SWITCH

Cut a circle ¼" in diameter out of conductive fabric or conductive fabric tape. Place the circle on one side of the poof. If you are using conductive fabric without an adhesive backing, glue the circle onto the poof. Sew the poof onto the top layer with the conductive fabric placed directly on top of the hole. Turn the top layer over. You should be able to see the conductive fabric.

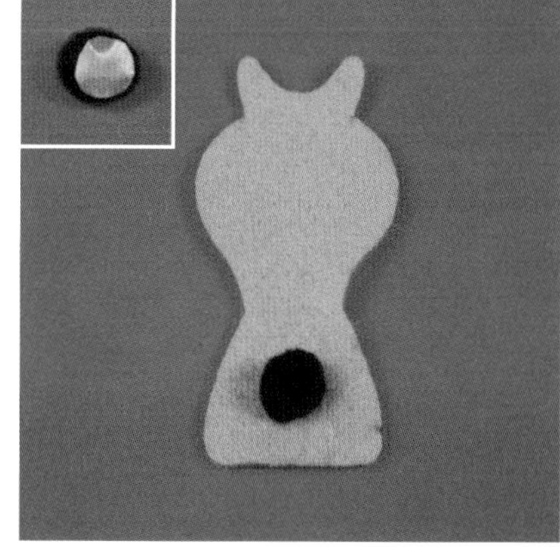

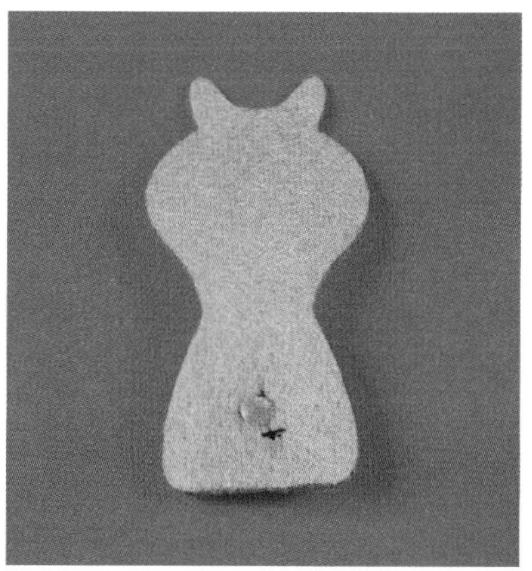

4. ASSEMBLE THE FINGER PUPPET

a. Using different sizes of googly eyes, personalize your character.

b. Place the middle, circuit layer right side up on top of the back layer. Using a utility knife, slice the back layer near the 2 snaps. Pull the snaps through the sliced holes in the back.

c. Align the edges of all 3 felt layers together. Using embroidery thread, stitch all 3 layers together, leaving the upper portion of the puppet open.

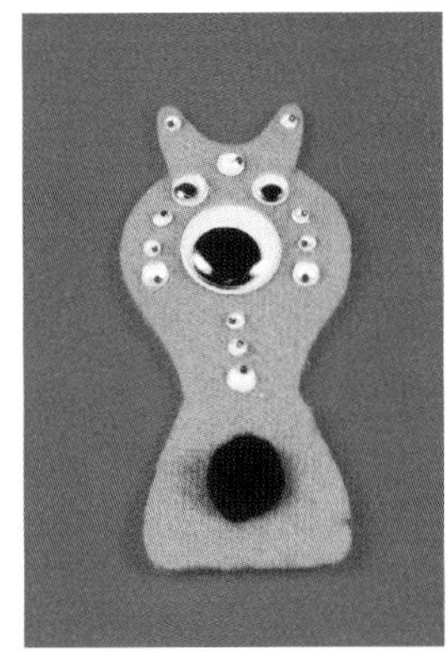

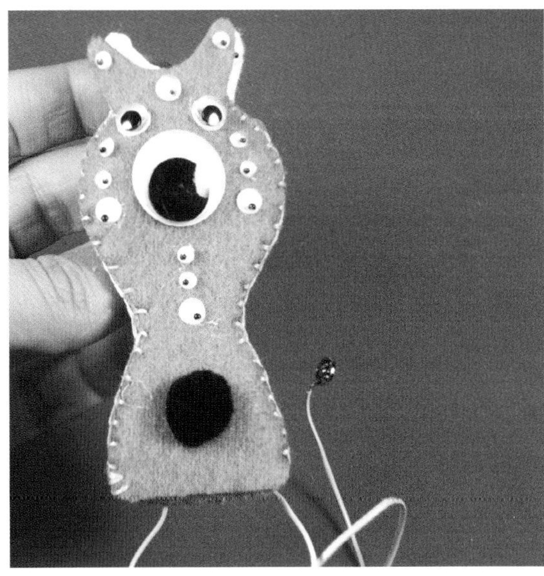

d. Fill the puppet between the back and middle layer with stuffing. Stitch the top of the puppet closed.

Congratulations! You're a third of the way there. Your first puppet is complete.

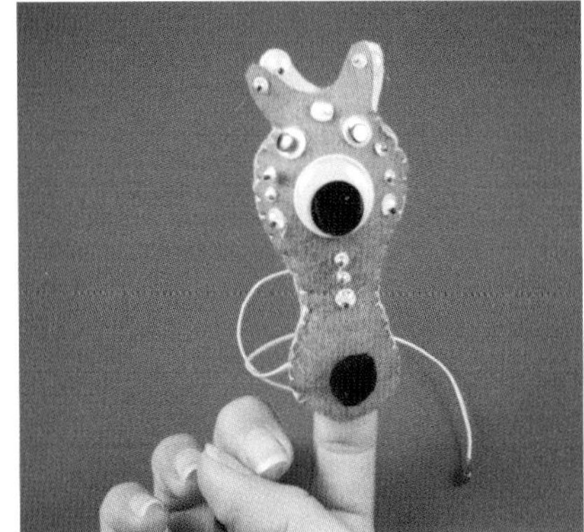

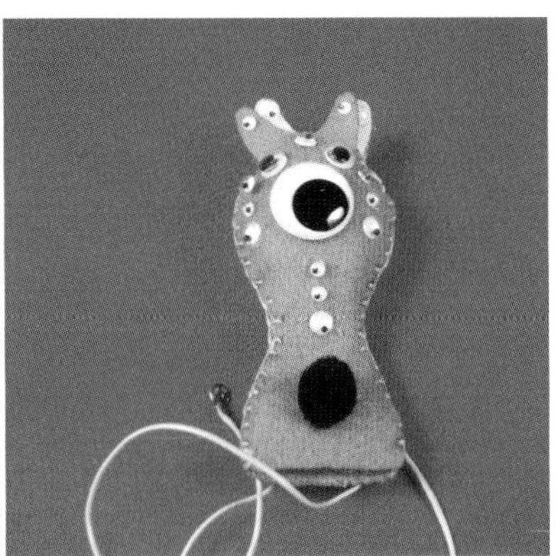

START » **BLUSH:** Puppet with LEDs

5. SEW THE CIRCUIT PATH

Blush is constructed in a similar fashion as Tremble, but with a slightly different circuit arrangement. Blush uses two 3mm red LEDs that must be sewn together in parallel. Refer to pages 22–23 to learn about wiring LEDs. Repeat Step 1 to cut the template and punch out the center hole and Step 3 to add the conductive fabric to the poof for Blush's soft switch. Make sure to only cut out the dotted line part of Template C on one layer, the top layer.

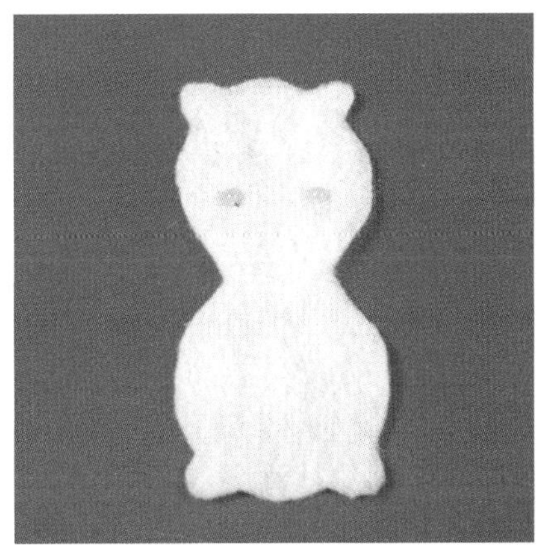

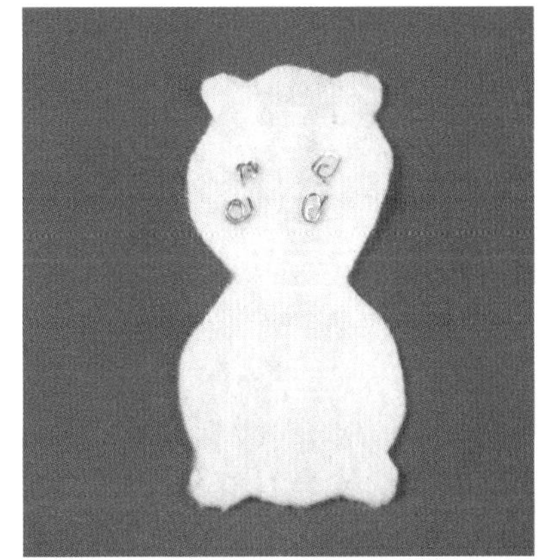

a. Slip the leads of an LED through the middle, circuit layer with the negative lead on top. Using needlenose pliers, curl the leads of the LED flush to the fabric. Repeat for the second LED.

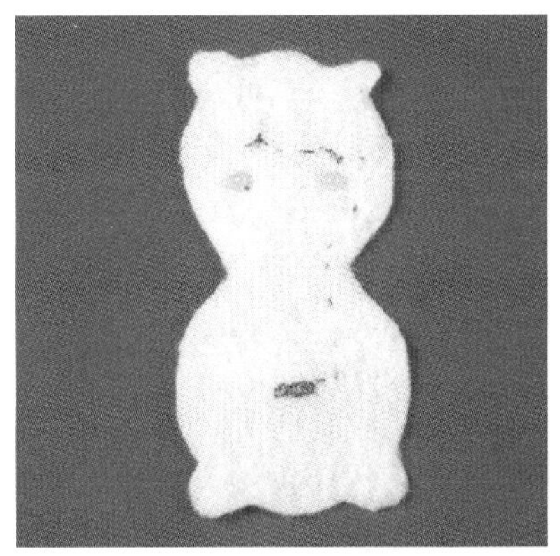

b. Using a sewing needle and conductive thread, sew a path from the top right, negative LED lead to the other. Continue sewing a path in a straight line toward the center switch mark. Once you reach the top of the mark, create the first contact point for the soft switch by embroidering a small patch of conductive thread. Knot and cut the thread.

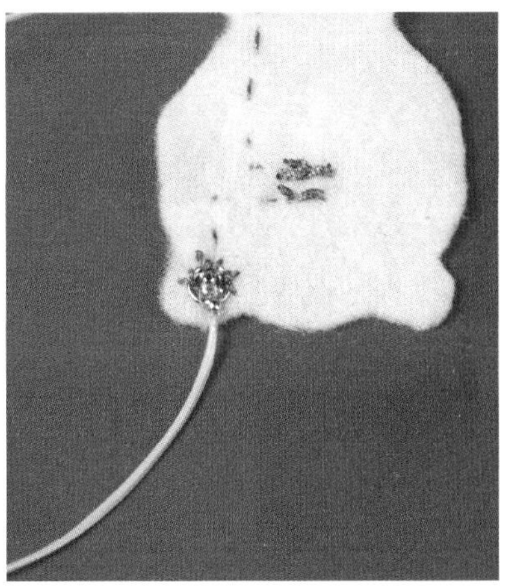

c. Again, using a sewing needle and conductive thread, create the second contact point for the soft switch by embroidering another small patch of conductive thread below the first. *These 2 patches of conductive thread should not touch.* Continue to sew a path to the lower corner about ¼" from the bottom and side of the felt. Sew the male snap end of the wire securely on the *back* of the felt.

d. Using conductive thread, sew a path from the bottom left, positive LED lead to the other. Continue to sew a straight line toward the lower corner, about ¼" from the bottom and side of the felt. Sew the female snap end of the second wire securely on the *back* of the felt (the side without LEDs).

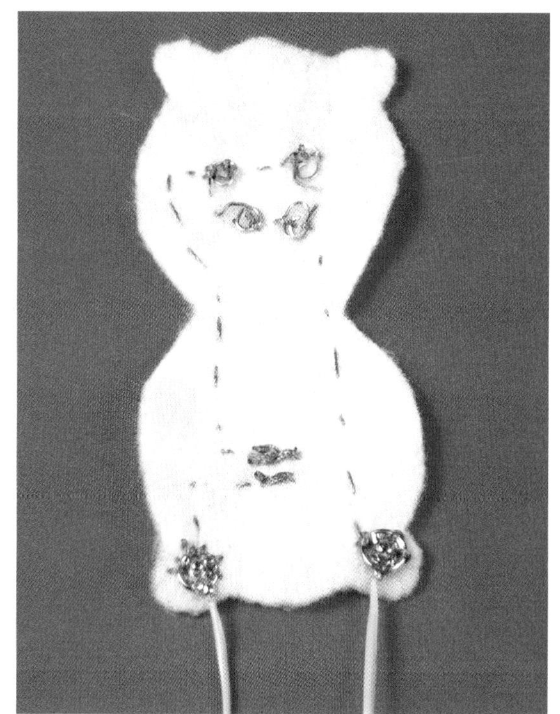

6. ASSEMBLE THE FINGER PUPPET

a. Using a set of googly eyes and embroidery thread, personalize your character. Blush, unlike Tremble, uses an extra layer of felt to design facial features and expressions.

b. Place the middle, circuit layer right side up on top of the back layer. Using a utility knife, slice the back layer near the 2 snaps. Pull the snaps through the sliced holes to the back.

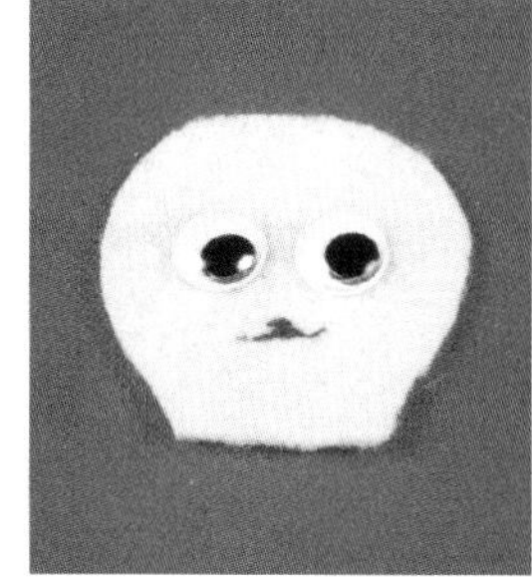

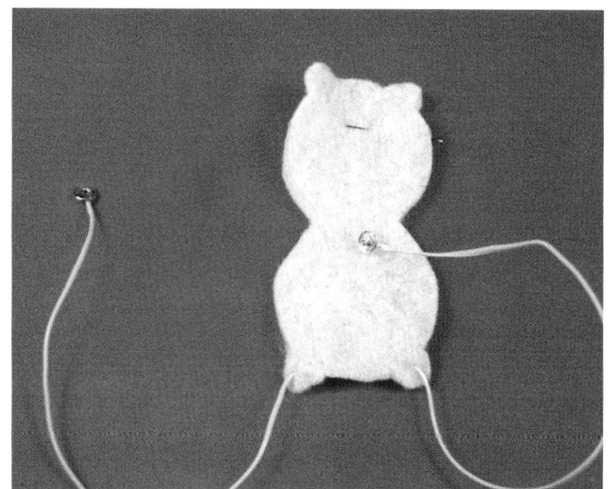

c. Align the edges of all 4 felt layers together. Using embroidery thread, stitch all the layers together, leaving the upper portion of the puppet open. Then fill the puppet between the back and middle layer with stuffing. Stitch the top of the puppet closed.

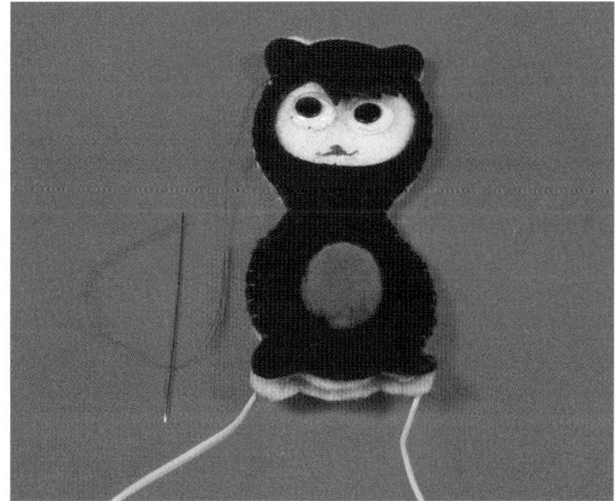

START » SPITFIRE: Puppet with Fiber Optics

7. ADD FIBER OPTICS TO AN LED

Almost there! For the last finger puppet, Spitfire, you will modify an LED by adding fiber optics.

a. Take a bunch of fiber optics and cut them to about 2" in length. Secure them together using a rubber band.

b. Add a touch of hot glue to the top of an LED. Quickly place the fiber optics at the end of the LED, and hold in place for 10–15 seconds or until the fiber optics adhere to the LED casing.

c. Remove the rubber band. Add a touch of hot glue to the bottom of the fiber optics and, while the hot glue is still warm, spread the fiber optics apart.

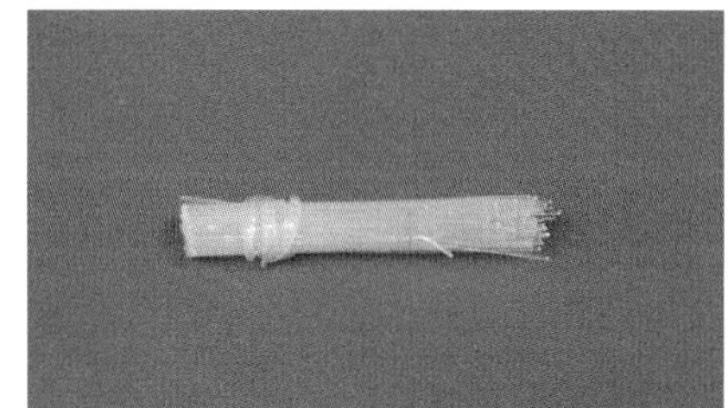

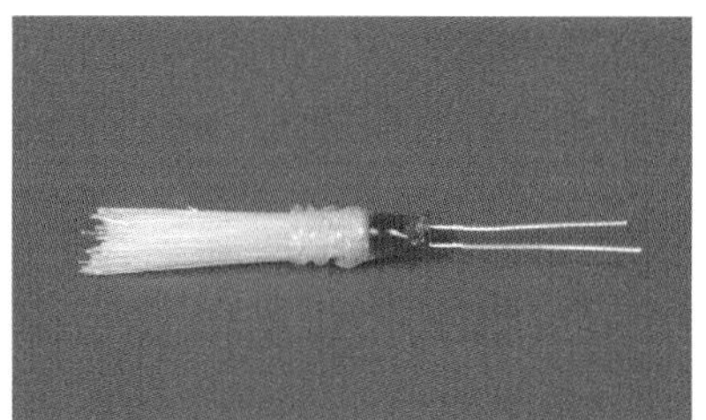

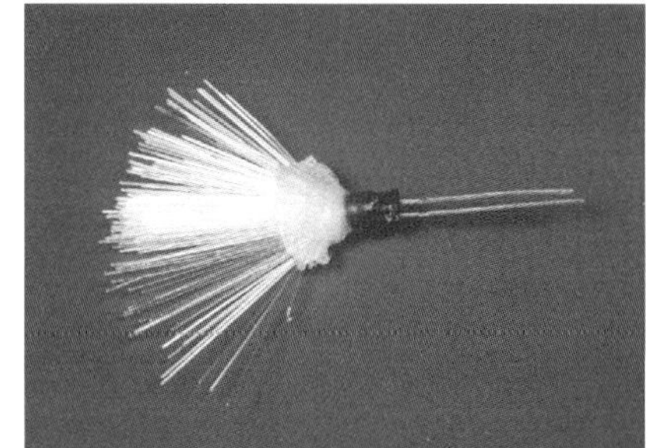

8. SEW THE CIRCUIT PATH

Once again, repeat Step 1 to cut Spitfire's template and punch out the center hole.

a. Using a utility knife, create a ½" slit in the upper portion of the top felt layer. Slip the fiber optics through the slit.

b. Grab the middle circuit layer and align the edges with the top layer. Pin the 2 layers together. Slip the leads of the LED through the middle circuit layer. Using needlenose pliers, curl the leads of the LED flush to the fabric, taking note of which is the short, negative lead and which the long, positive lead. Place the negative LED lead facing the left and the positive facing the right.

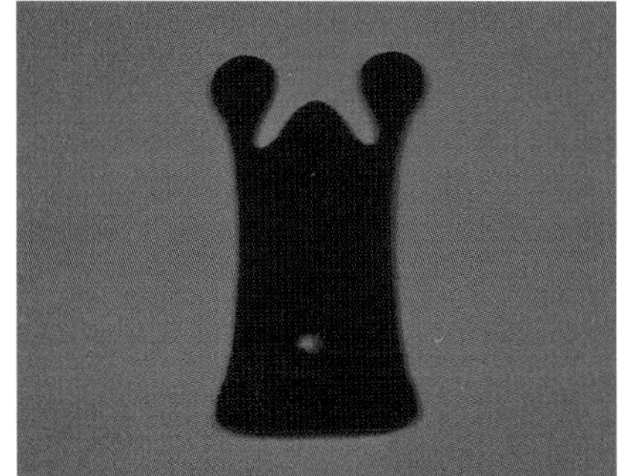

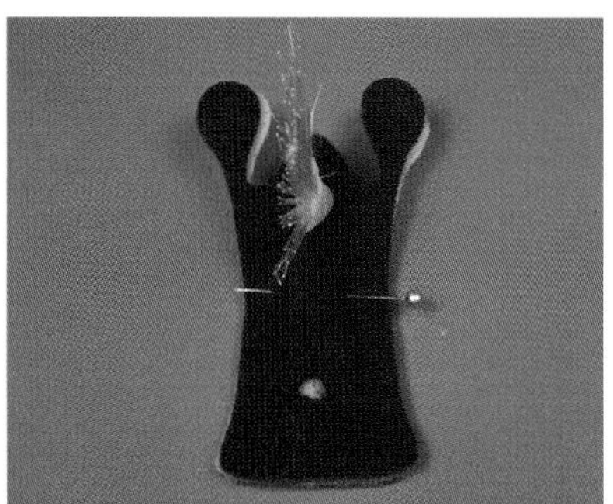

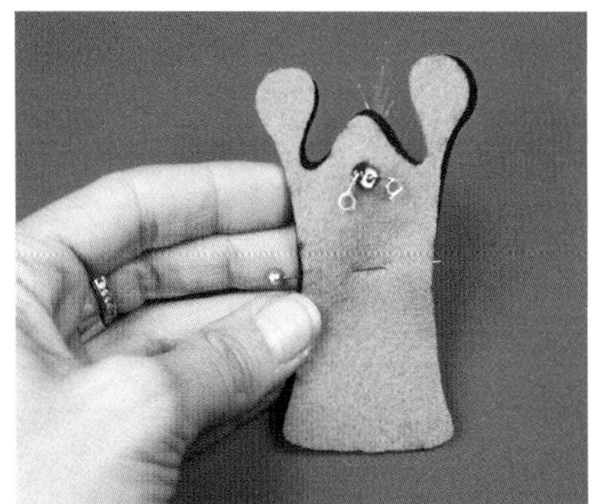

c. Using a sewing needle and conductive thread, sew a path from the negative LED lead toward the center switch mark. Once you have reached the top of the mark, create the first contact point for the soft switch by embroidering a small patch of conductive thread. Knot and cut the thread.

d. Using conductive thread, create the second contact point for the soft switch by embroidering another small patch of conductive thread below the first. *These 2 patches of conductive thread should not touch.* Continue to sew a path to the lower corner, about ¼" from the bottom and side of the felt. Sew the male snap end of the wire securely onto the *back* of the felt. Make sure all sewing only goes through the middle layer.

e. Using conductive thread, sew a path from the positive LED lead toward the lower corner about ¼" from the bottom and side of the felt. Sew the female snap end of the second wire securely onto the *back* of the felt.

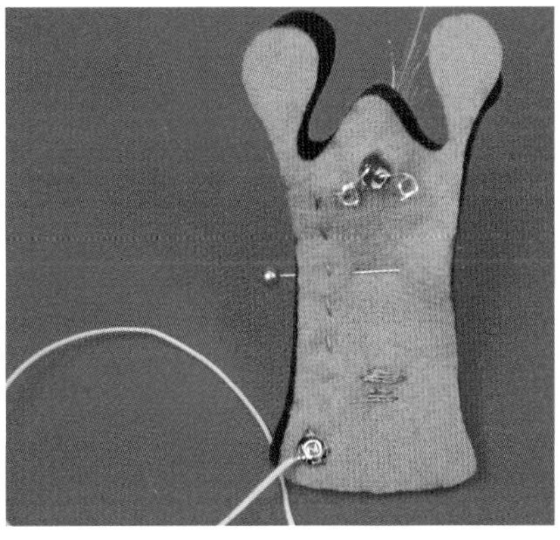

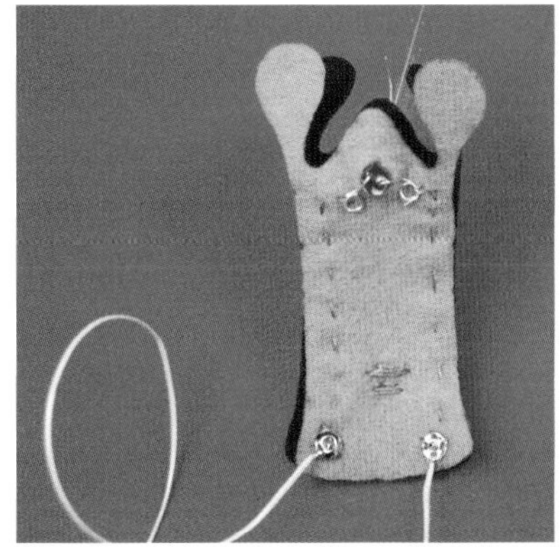

NOTE: Make sure you only sew through the middle circuit layer. You don't want the conductive thread visible on the front layer.

f. Sew the poof with the conductive fabric onto the front fabric layer, and personalize your finger puppet with googly eyes and embroidery thread. Finally, hand-stitch all 3 layers together.

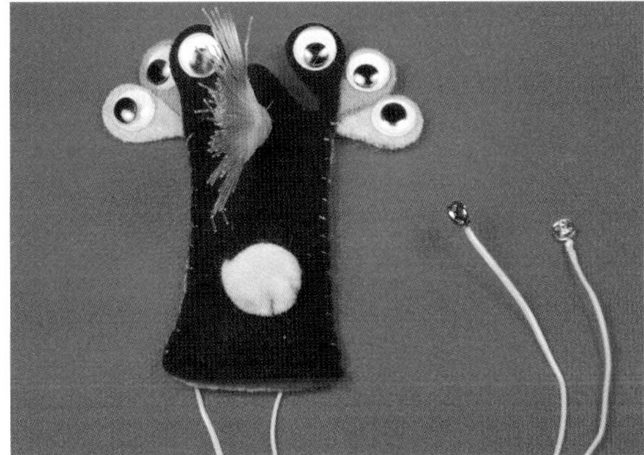

9. MAKE THE POWER SOURCE BRACELET

a. Different battery holders come with different wires. Cut the wires to 2".

b. Using a marking pen, trace Template E onto the felt twice. Then cut out the 2 pieces.

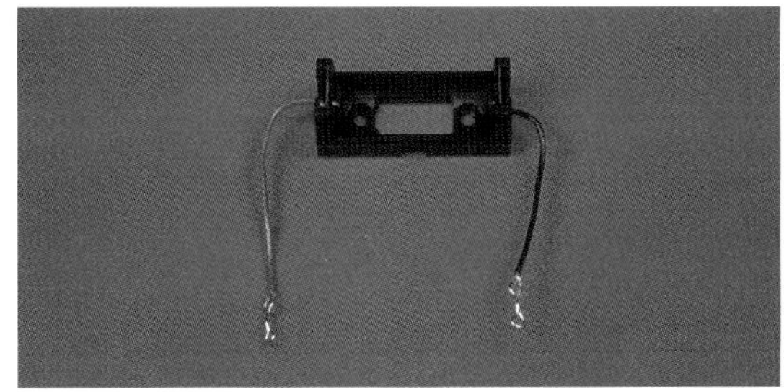

c. Using conductive thread, sew a straight line 4" in length, 1½" from the side, and ¾" from the top on the top bracelet layer. Using conductive thread, sew another straight line 4" in length, 1½" from the side and ¾" from the bottom.

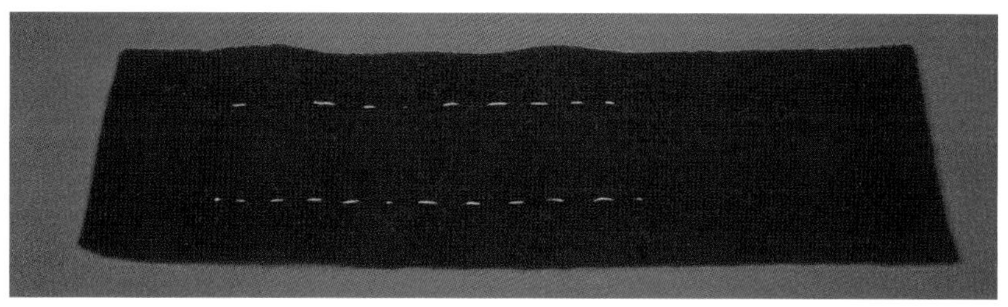

d. Place the battery holder 2" from the right side of the top layer, with the positive, red lead on top. Using a utility knife, make 2 small slits in the fabric. Slip the leads of the battery holder through to the opposite side. Using conductive thread, sew the loop of the positive lead to the top stitched conductive path. This will be your positive conductive path. Repeat for the negative loop, sewing it to the bottom stitched conductive path. This will be your negative conductive path.

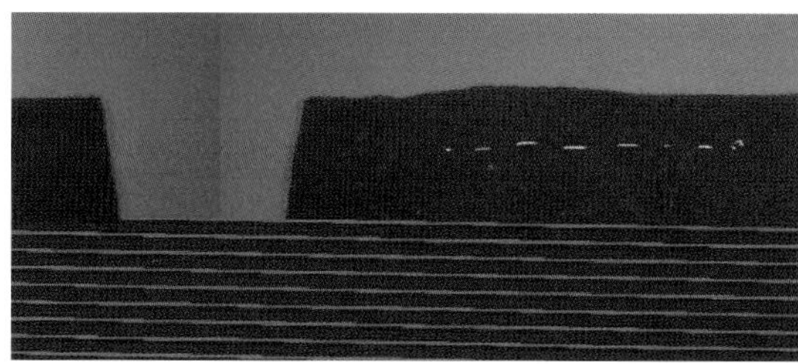

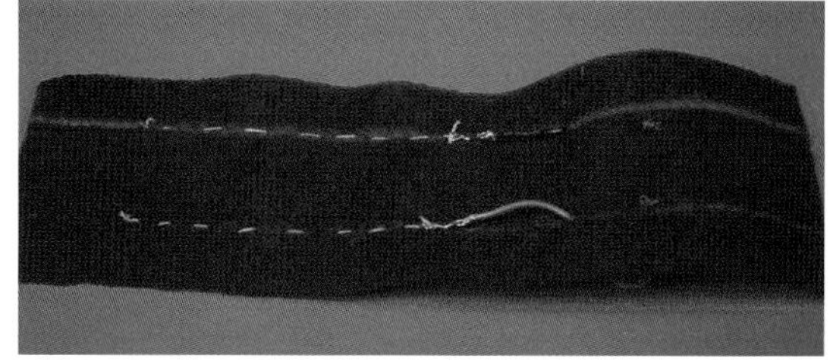

e. Take 3 snaps, and using conductive thread, sew each female snap individually onto the positive conductive path. Repeat for the male snaps, sewing them to the negative conductive paths.

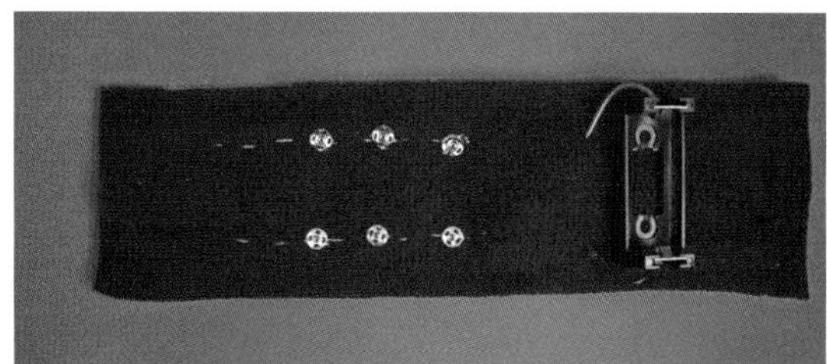

f. Aligning the edges of the top and bottom felt layers, machine-sew the 2 layers together. Next, create a buttonhole on the right side of the bracelet. Sew the button onto the left side.

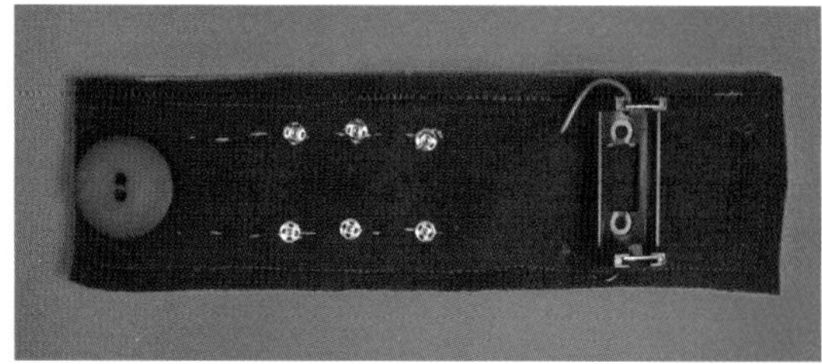

g. Put the bracelet on and snap the wires from the finger puppet into place. Slip a battery in the battery holder and play. Press each puppet's belly button and bring your puppets to life!

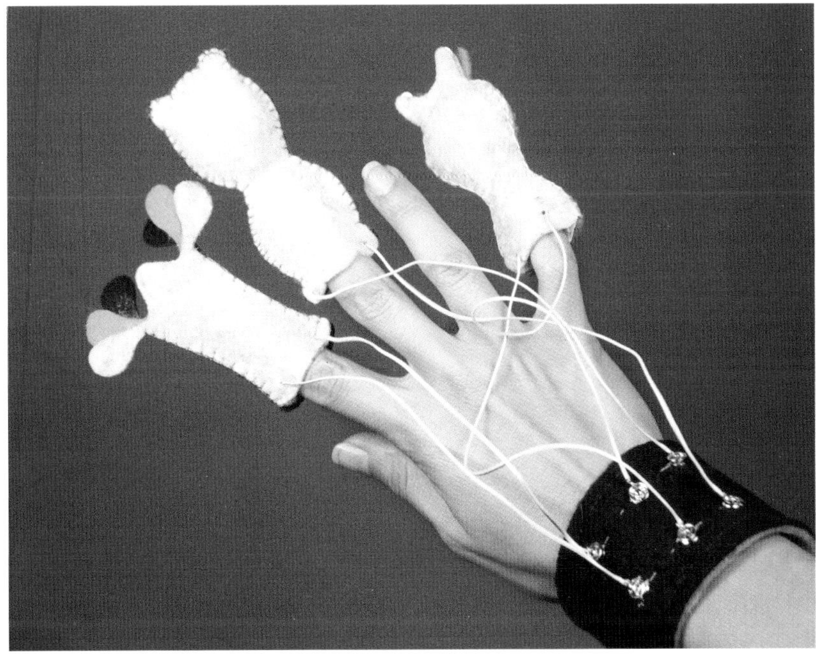

The whole gang.

GLAM THE GLO BUG

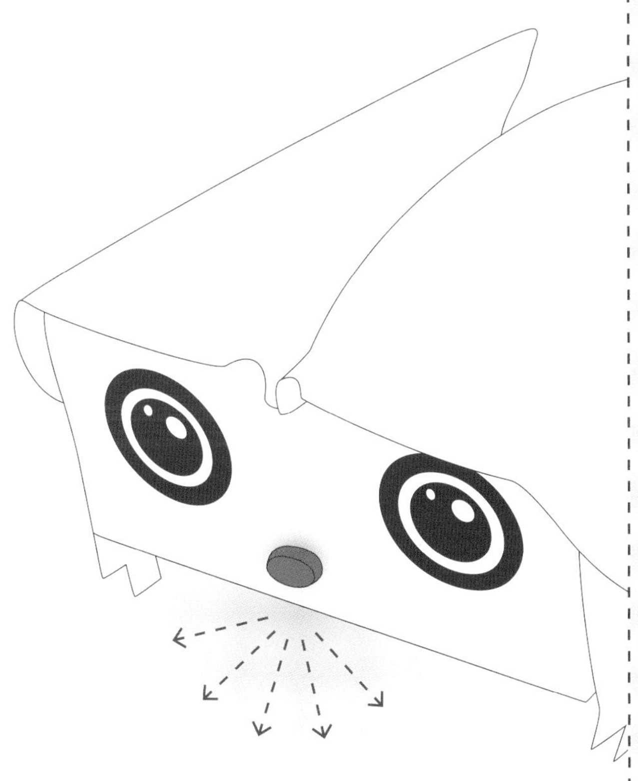

Glam is a marvelous, exquisite little bug that radiates a warm, gentle glow. We're not sure why, but it is rumored that he glitters when he dreams.

HOW IT WORKS »

Glam's central nervous system uses a mini photocell, also known as a light-dependent resistor (LDR), to sense changes in the level of ambient light. Glam's transistor is his internal switch, which turns the LED on and off. The transistor, in turn, is turned on or off by the photocell. In bright light, the photocell's resistance is low, so the transistor turns off, which turns the LED off. In low light, the photocell's high level of resistance switches the transistor on, and Glam begins to glow. To adjust Glam's sensitivity to light, the 50K trimpot is used to fine-tune the level of darkness required before the LED lights up.

COMPLEXITY: INTERMEDIATE

WHAT YOU WILL LEARN

- Work with LEDs
- Build a circuit with conductive tape (or ink)
- Work with photocells and transistors

RELATED TUTORIALS

- Building a Simple Circuit (27-33)

MATERIALS

- **8½"×11" sheet of colored paper** preferably a light-colored heavier stock
- **Conductive fabric tape, 12"** Ni/Cu fabric tape from lessemf.com
- **Double-sided tape**

TOOLS

- Computer
- Printer
- Scissors
- Utility knife
- Needlenose pliers
- Sewing needle

ELECTRONICS

- **BC548 NPN transistor** part #254781 jameco.com
- **Mini photocell** part #202403 jameco.com
- **1KΩ resistor**
- **470Ω resistor**
- **LEDs (2)** preferably white
- **50K trimpot**
- **9V battery connector**
- **9V battery**

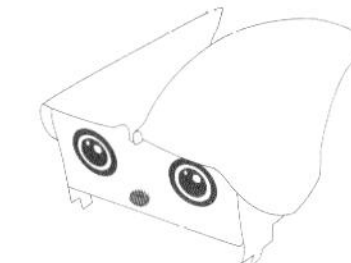

TEMPLATES

Templates and circuit diagram available at fashioningtechnology.com/globug.

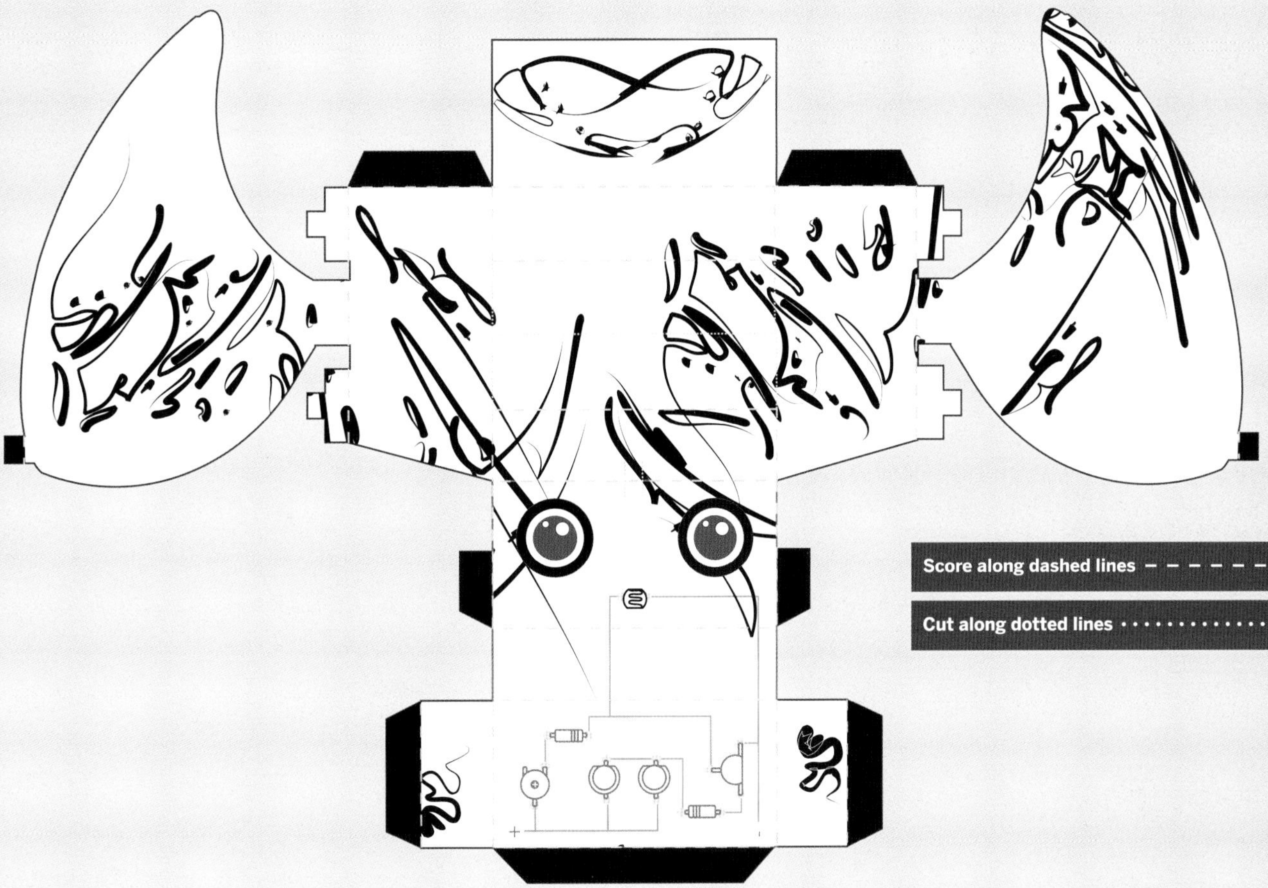

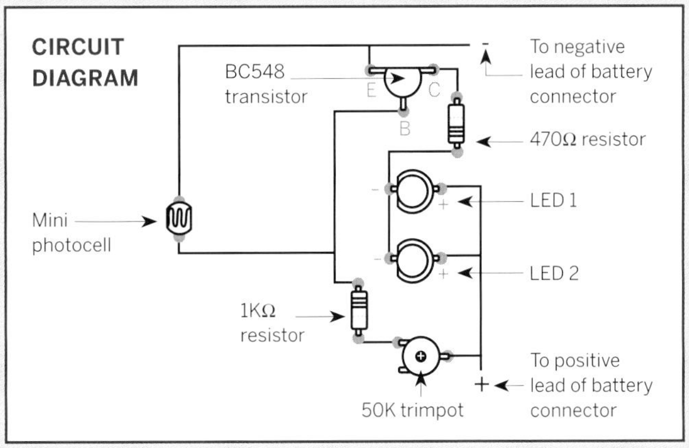

START »

1. CUSTOMIZE TEMPLATE

Two templates designed to print on 8½"×11" sheets of paper are provided at fashioningtechnology.com/globug: one blank and the other with illustrated patterns. To personalize Glam with a custom skin, bring the templates into any graphics software and create a custom pattern. Print and cut out the body and circuit templates. Heavy matte paper is suggested for sturdiness.

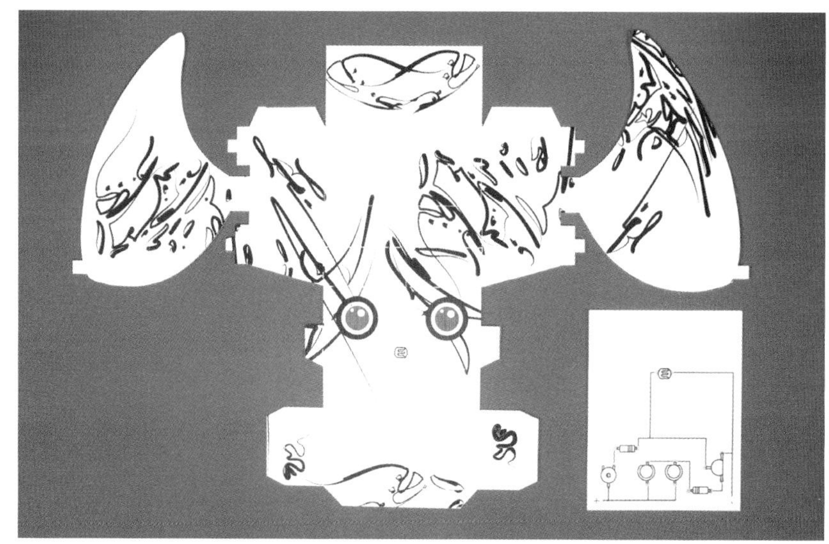

2. MAKE THE CIRCUIT

a. Using scissors or a utility blade, cut the conductive fabric tape into thin 1"–2" long sections. You will be using thin sections of conductive fabric tape to create a path for the electricity to flow through the circuit. Secure the tape directly onto the solid lines of the circuit template, making sure that the sections of tape perpendicular to each other overlap.

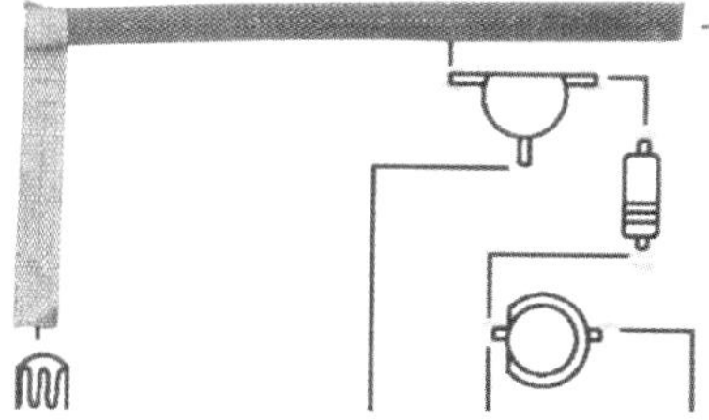

Use a Conductive Pen

» You can swap conductive fabric tape with conductive paint/pen to create the conductive paths on the template. Using a conductive pen or paint, trace directly over the solid lines. To adhere the components in place, you will need to use conductive epoxy.

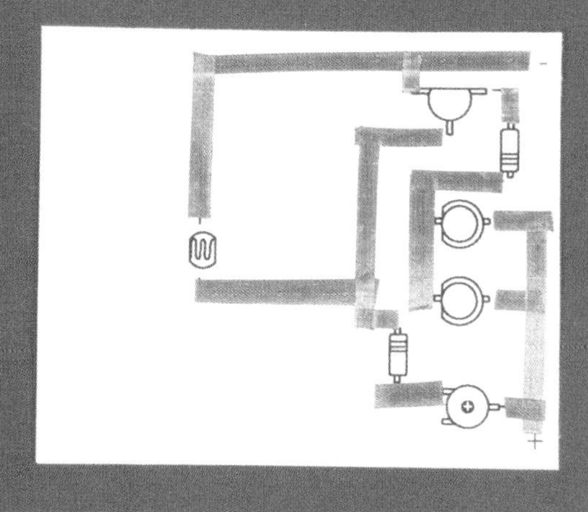

b. Using needlenose pliers, gently bend the leads of the components at 90° angles. Starting with the transistor, position it in place as illustrated on top of the circuit. Trim the leads to prevent them from touching other components and causing a short. You can use a small dab of hot glue to position the transistor into place, but be sure not to coat the leads with hot glue.

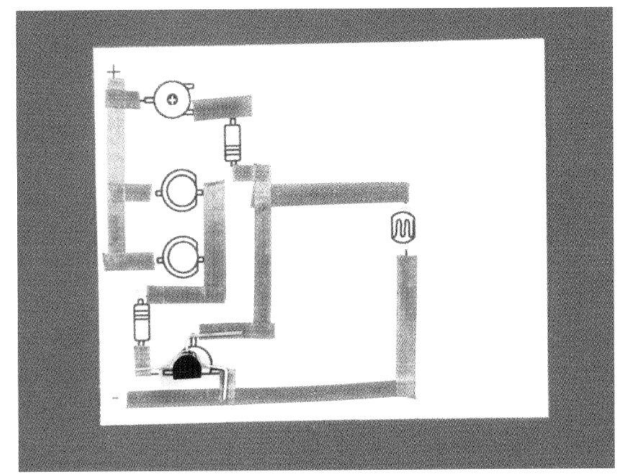

c. Place conductive fabric tape directly on top of the transistor leads, securing the transistor in place on top of its corresponding circuit path.

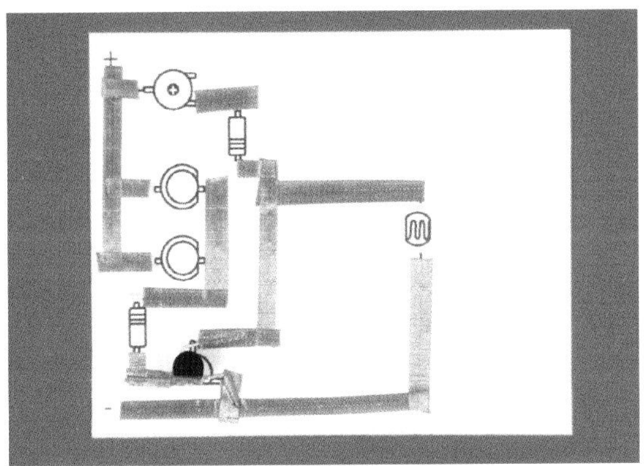

d. Repeat for the rest of the components (except for the photocell), taping each component into its corresponding place as illustrated on top of the circuit. Refer to the circuit diagram on page 177. *One lead of the 50K trimpot will NOT be connected to any other component.*

NOTE: The 470Ω resistor should be connected to the collector lead of the transistor (labeled C in the diagram on page 177) while the 1KΩ resistor should be connected to the base lead of the transistor (labeled B).

e. Place the circuit template face down. Pierce the leads of the photocell through the back side of the template, taping the leads onto their corresponding circuit paths. The photocell will now be positioned on the back of the template with its leads taped securely to the front.

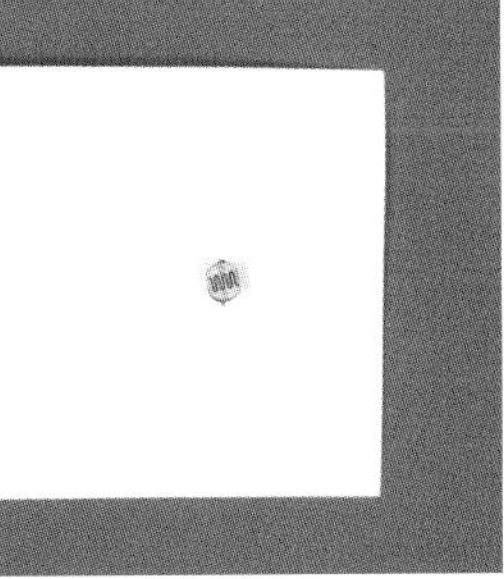

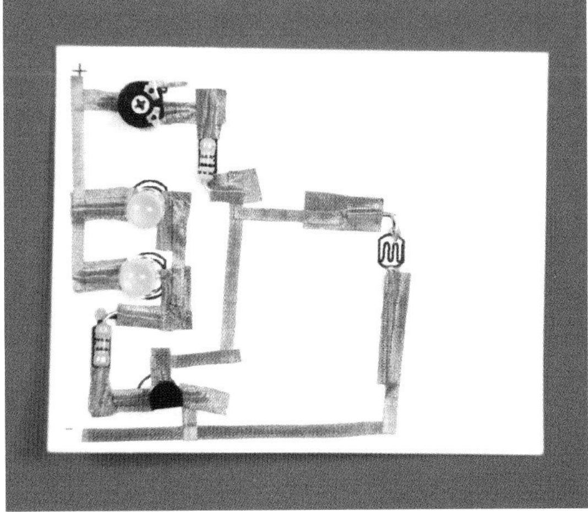

f. Using a utility knife, cut out the photocell illustration from the body template. Place the body template face down. Align the circuit template with the photocell facing up underneath Glam's face on the bottom of the body template. Slip the photocell through the hole cutout. Using double-sided tape, secure the circuit template onto the body template.

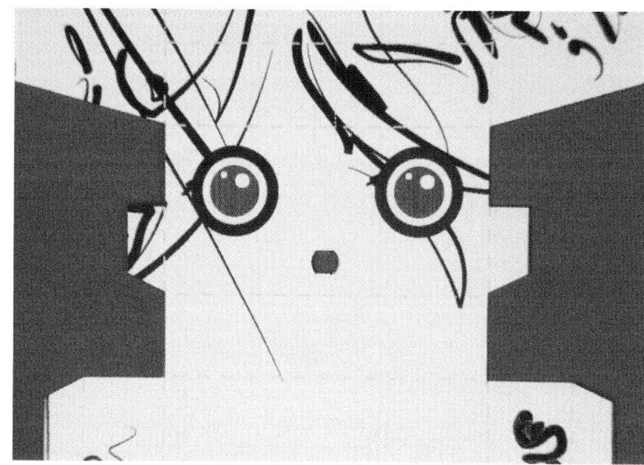

g. Using a needle, pierce a hole through the center of both the "+" and "−" symbols on the circuit template. Grab the battery connector, and from the top of the body template, slip the red positive wire through the "+" hole and secure it to the circuit path using conductive tape. Repeat for the black negative wire, slipping the wire through the "−" hole.

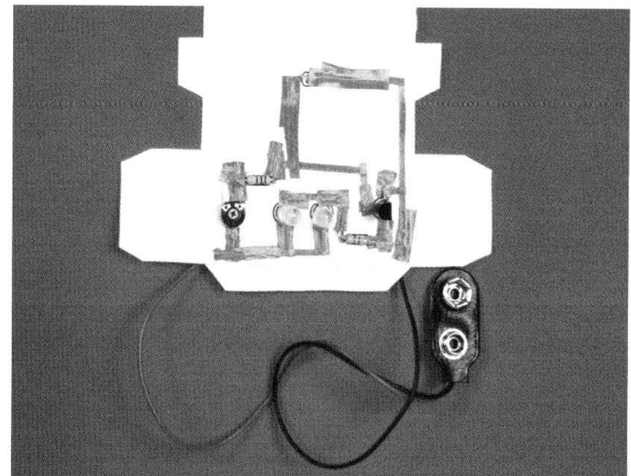

3. FOLD THE FORM

a. Using a utility knife, cut the middle section along the dotted lines and fold inward. The large dashed lines indicate score lines, and the small dotted lines indicate cut lines.

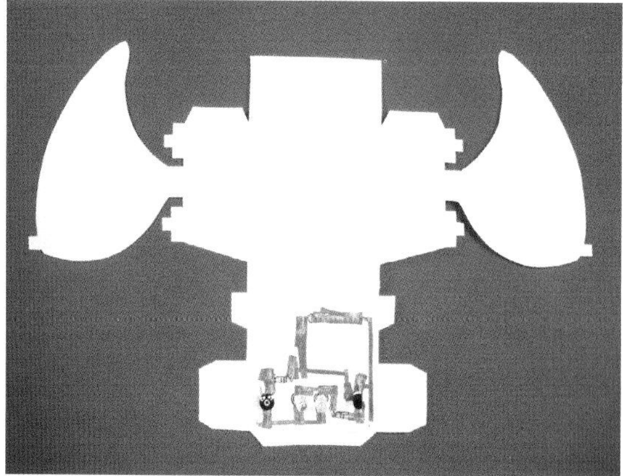

b. Fold the body along the designated score lines, following the instructions on the body template.

Use double-sided tape to adhere the fold flaps (colored in black on the template) to each other. Using a pair of scissors, snip the dotted lines between the eyes. Add a small piece of double-sided tape to the small fold tabs on the wings. Fold the wings upward and secure them into the corresponding slit.

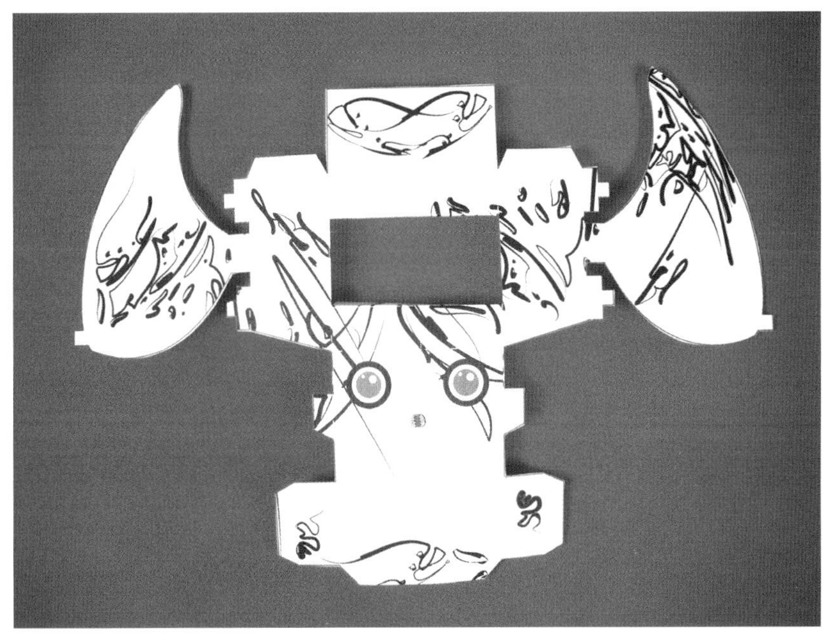

4. PLACE THE BATTERY

Turn the bug over and snap the battery into the battery connector. Gently slip the battery into the groove. Once the battery is in place, use tape to secure the wires onto the bottom of the bug. Turn the lights off, lulling Glam to sleep, and watch him glow as he begins to dream.

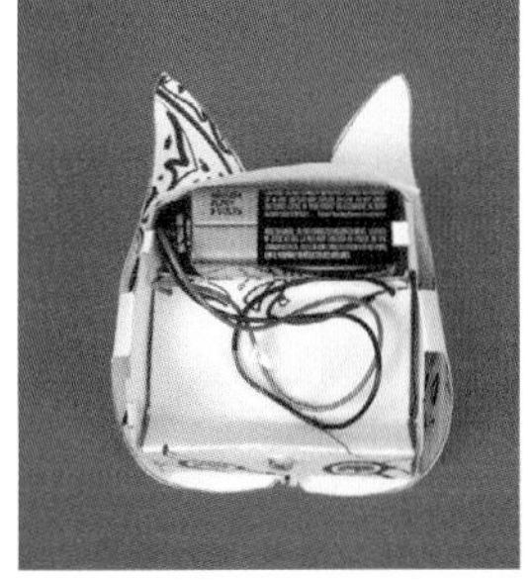

NOTE: Using a small screwdriver, you can change Glam's sensitivity to light by adjusting the 50K trimpot.

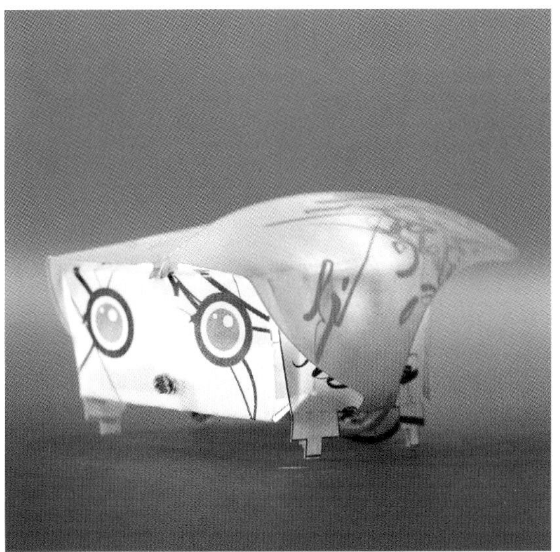

FINISH ⊠

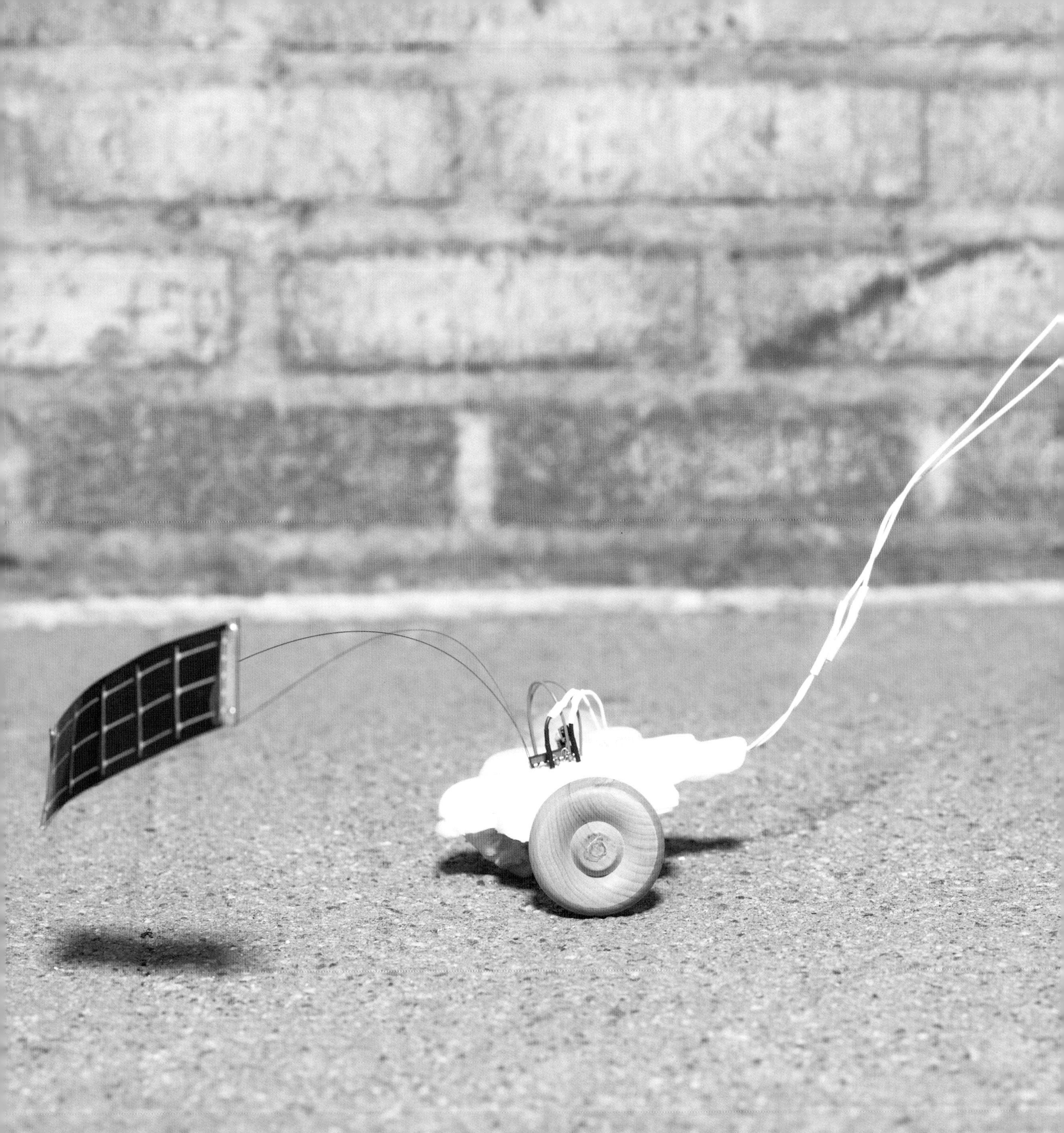

SOLAR CRAWLER

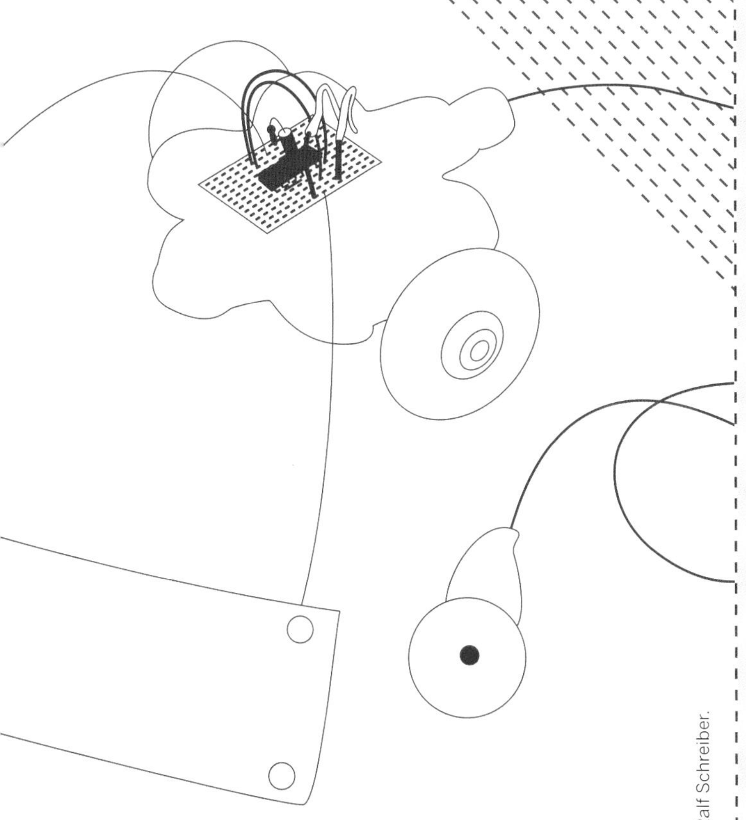

The circuit design is the creation of Ralf Schreiber.

The Solar Crawler thrives in sunlight, magically translating the sun's invisible rays into song. Located at the end of the Crawler's pull string is a small piezo speaker that emits sound. On a bright, sunny day, take the crawler for a stroll and discover the swishing sounds of the sun. Its unique sounds will fascinate both children and adults alike.

HOW IT WORKS »

The Solar Crawler uses a flexible solar panel suspended on piano wire to energize the circuit and produce variable sounds. The circuit is built on a breadboard to allow you to adjust and alter the sounds over time. The constant in the circuit is a Hex-Schmitt inverter IC, but the capacitors, resistors, and even the solar panel can be swapped to create varying sounds.

No two Solar Crawlers sound alike. Once the circuit is built, continue to experiment by swapping different-value capacitors and resistors until you find your individual crawler's unique voice. Build a swarm of them and listen to them sing in concert.

COMPLEXITY: ADVANCED

WHAT YOU WILL LEARN

- Work with solar cells
- Sculpt with Polymorph plastic
- Work with ICs
- Use a breadboard
- Solder

RELATED TUTORIALS

- Building a Simple Circuit (27-33)
- Measuring the Voltage of a Circuit (71)

MATERIALS

- **Decorative baking pan** minimum 1" deep
- **Wooden wheels (2)**
- **Wooden dowel** diameter to fit wheels
- **Aluminum tubing** The diameter must be large enough for the wooden dowel to rotate freely.
- **Polymorph pellets, 250 grams** This amount will depend on the size of your mold. My mold was 5¼" in diameter and 1¼" deep.
- **Piano wire, 24" strand, 22–24 AWG** Diameter must be small enough to plug into breadboard.
- **Solid wire, 9'**
- **Jumper wire, 4" lengths (3)**
- **Variety of jumper wires or solid wire**
- **4" heat-shrink tubing, ⅛" diameter**
- **Solder**
- **Masking tape**
- **Red permanent marker**

NOTE: Quantities of wheels, dowels, and aluminum tubing listed here should be doubled to make a 4-wheeled crawler.

TOOLS

- Tube cutter
- Soldering iron
- Scissors
- Needlenose pliers
- Wire cutters
- Wire strippers
- Wooden or plastic tongs
- Utility knife
- Ruler
- Hot glue gun
- Heat gun or hair dryer

ELECTRONICS

- **Solderless breadboard** part #700-00012 parallax.com
- **IC Hex-Schmitt inverter, 74HC14** part #45364 jameco.com
- **Capacitors ranging from 1μF to 47μF (2)**
- **Resistors ranging from 1KΩ to 100KΩ (3)**
- **Piezo speaker** Allelectronics.com has a good selection.
- **Flexible solar cell with a minimum voltage > 2.5V**

START »

1. CREATE THE CHASSIS

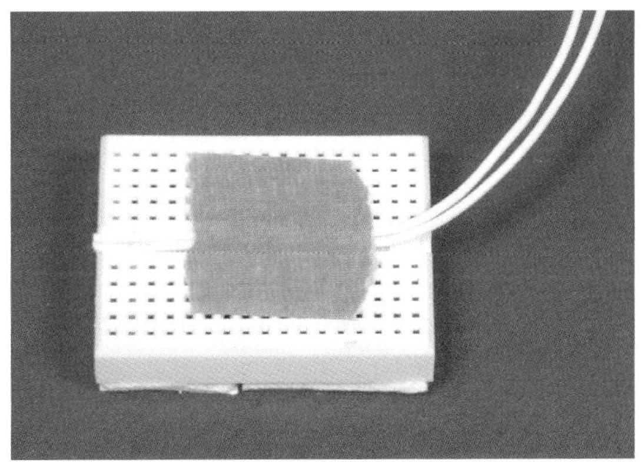

a. Using wire cutters, cut the 9' solid wire in half, into two 4½' pieces. Using wire strippers, strip about ¼" of insulation off both ends of the 2 lengths of wire. Place the wires in the center groove of the breadboard. The tips of the wires should align with the edge of the breadboard. Using masking tape, secure the wires in place.

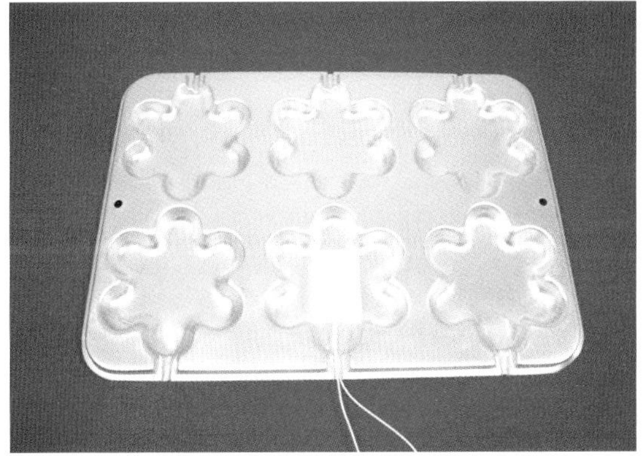

b. Place the breadboard and wire assembly face down in the center of the baking pan. To determine the length of the wooden dowel axle, use a ruler and measure the pan's diameter at the location where you plan on placing the axle and also the width of the wooden wheel. The best axle placement depends on the design of the pan. The length of the wooden axle should be approximately the length of the pan's diameter plus the width of the 2 wheels plus 1". The extra ½" on each side of the wheel will allow for smooth rotation.

Once you have determined the axle length, use a utility knife to cut the wooden dowel to the appropriate length. Then use a tube cutter to cut the aluminum tubing to a length 1" shorter than the wooden axle.

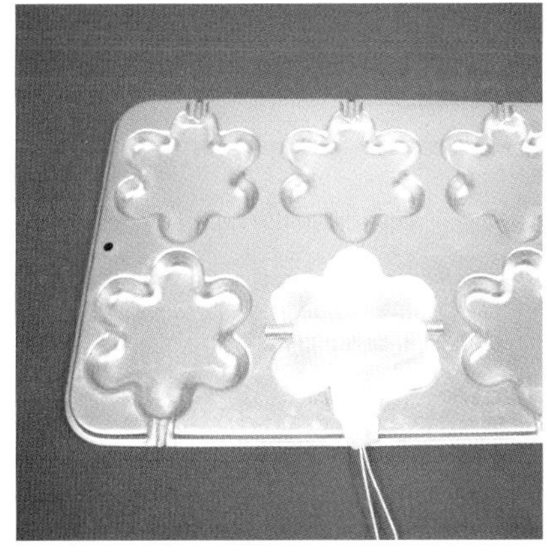

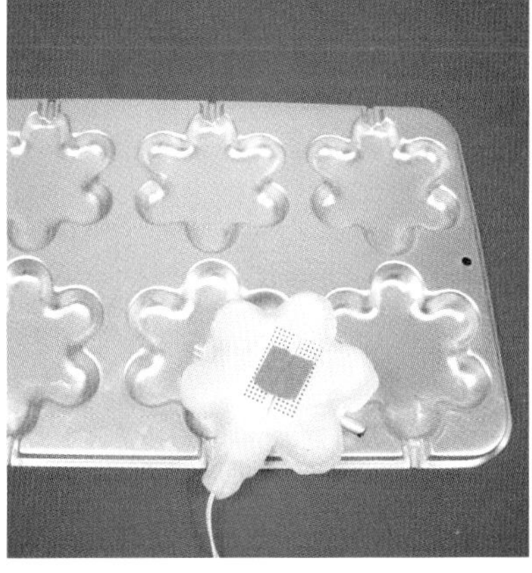

c. Heat the Polymorph plastic according to the instructions on the package until it is malleable. The Polymorph should be transparent at that time. If you use hot water to heat it, using a pair of tongs, carefully remove the Polymorph from the hot water and squeeze out the excess water. Quickly fill the pan around the breaboard with the Polymorph, embedding the breadboard and the wires in the Polymorph.

d. Place the aluminium tubing at the axle location and secure it in place with Polymorph.

Then slip the Polymorph-breadboard assembly out of the mold. This will be the chassis for the Crawler.

e. Place one of the wheels on one end of the axle and slip the axle through the aluminum tubing. Place the second wheel on the axle. Add a touch of hot glue to both wheels, securing them to the axle.

NOTE: The axle should be able to rotate freely in the aluminum tubing, allowing the axle and wheels to spin. If the axle can't spin easily, use aluminum tubing with a larger diameter.

2. BUILD THE CIRCUIT

a. Remove the wires from the center groove of the breadboard; these will be the speaker wires. Using wire cutters and strippers, cut a 4" jumper cable in half and strip about ¼" of insulation off both ends.

b. Using scissors, cut two 1" pieces of heat-shrink tubing. Slip a piece of heat-shrink over each speaker wire near the breadboard.

Twist together the stripped end of the jumper wire and the speaker wire. Solder. Slip the heat-shrink over the exposed wires and, using a heat gun or hair dryer, shrink the tubing until it is snug. Repeat for the second wire. Using a red marker, mark both ends of one of the speaker wires as the positive wire, as pictured.

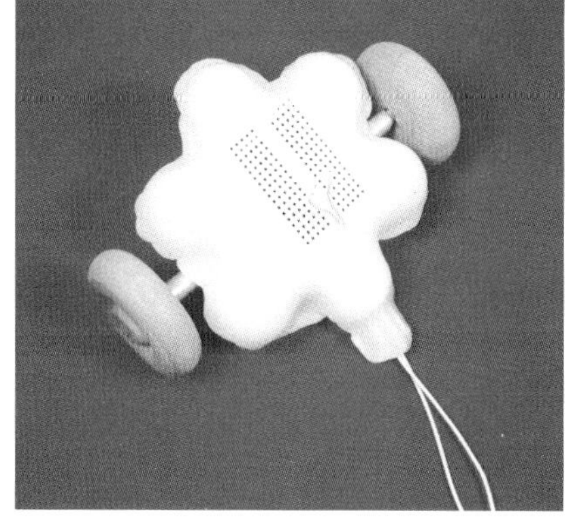

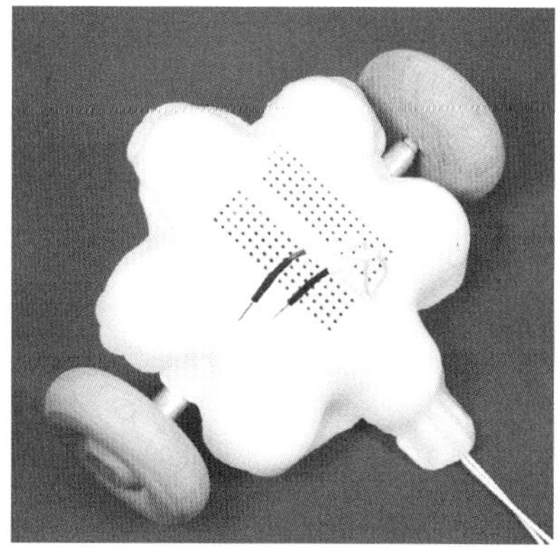

NOTE: This circuit is a modification of the Aerial the Birdie Brooch circuit. Please refer to the circuit diagram on page 113.

c. Place the IC 74HC14 chip in the middle of the board along the divider line so that half the leads are on one side and the other half are on the other side. Position the IC so the half circle notch is facing away from the speaker wires. The upper left pin will be pin 1. Using jumper wire, connect pin 1 to pin 4. Using another jumper wire, connect pin 8 to pin 11.

NOTE: If you are unfamiliar with using a breadboard, refer to page 30 for a detailed explanation.

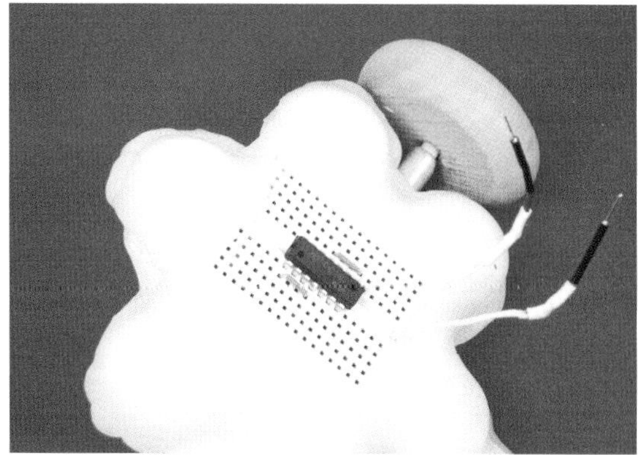

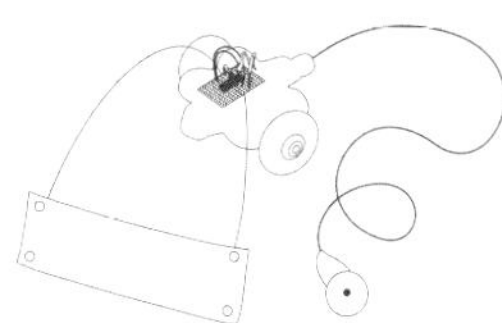

d. Using wire cutters, cut the leads of resistors A-C (two 10KΩ and one 100Ω) and capacitors B-C (10μF) shorter so that they can fit flush to the board. Connect capacitor C (10μF) to pins 9 and 10. Polarity doesn't matter in this circuit. Connect resistor C (100Ω) between pins 8 and 9.

NOTE: The Aerial Birdie Brooch circuit on page 113 has been modified: capacitor A has been eliminated from the design of this circuit.

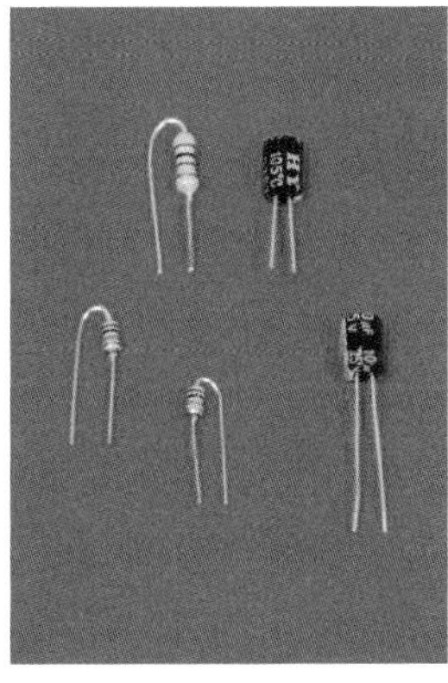

e. Connect one lead of capacitor B (10μF) to pin 5 and the second lead to a free hole on the breadboard, ideally to a row next to pin 7 of the IC. Polarity doesn't matter in this circuit design. Then connect the remaining 2 resistors, A and B (both 10KΩ) to pins 3 and 4, and 5 and 6.

f. Using the 4" jumper cable, connect pin 6 to pin 13. Using another 4" jumper cable, connect pin 12 to the second lead of capacitor B (the one not connected to pin 5). Then connect the negative speaker wire to pin 7 and the positive speaker wire (red) to pin 3.

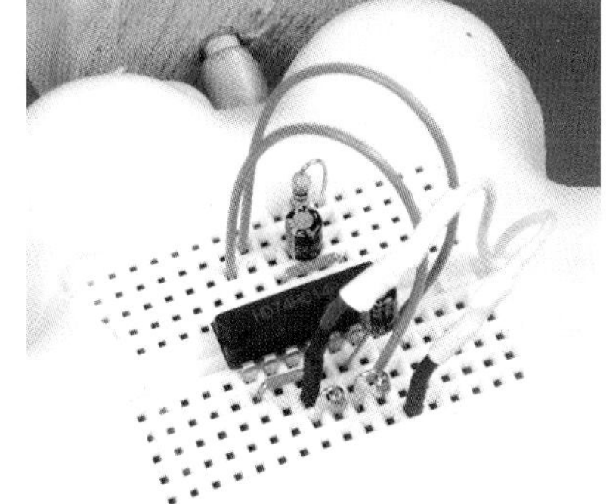

g. Solder the piano wire to the contact points on the solar panel. Cut the piano wire in half into two 12" lengths. Using a red marker, mark the positive contact. If you don't know which is the positive contact, use a multimeter to distinguish the positive contact from the negative. Refer to page 71 to learn more on how to determine polarity using a multimeter.

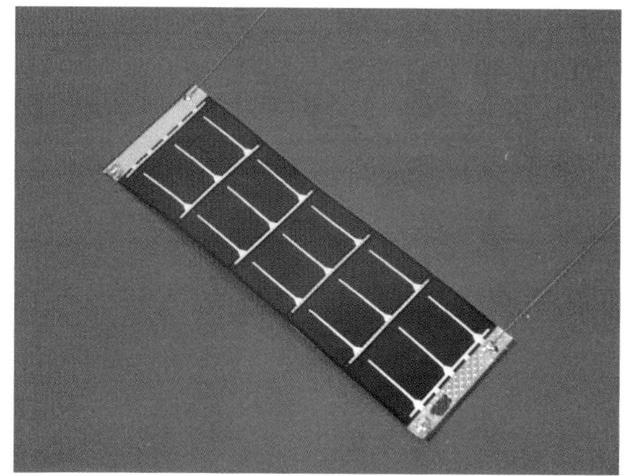

h. Gently bend the piano wires so that they wrap to the back side of the solar panel.

Connect the positive piano wire (soldered to the positive solar panel contact) to pin 14 and the negative to pin 7.

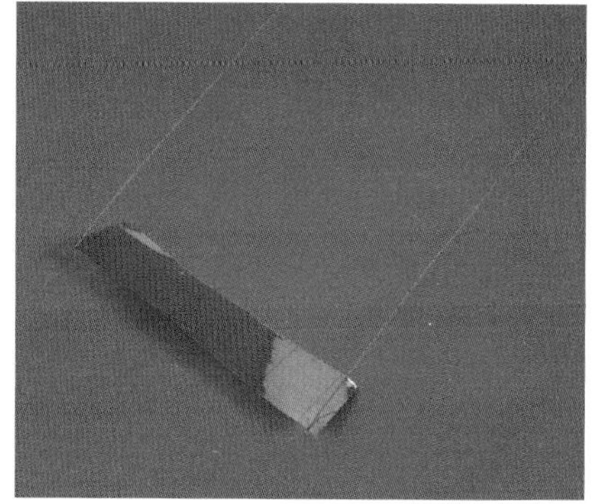

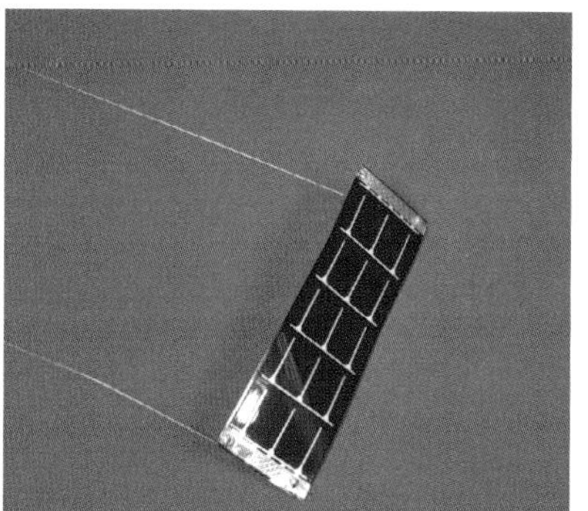

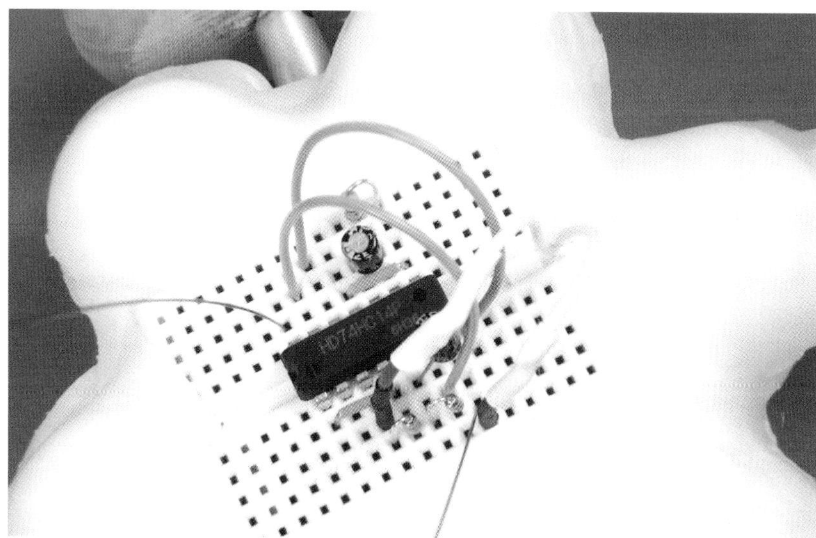

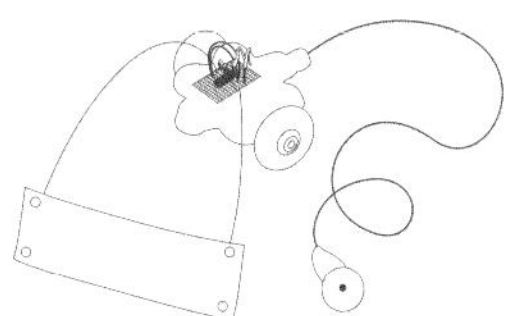

3. CREATE THE SPEAKER PULL STRING

a. Using scissors, cut two 1" pieces of heat-shrink tubing. Slip a piece of heat-shrink over each speaker wire. Twist together the positive speaker wire extending from the chassis with the positive wire extending from the speaker. Solder. Slip the heat-shrink over the exposed wires and, using a heat gun or hair dryer, shrink the tubing until it is snug. Repeat for the negative wire.

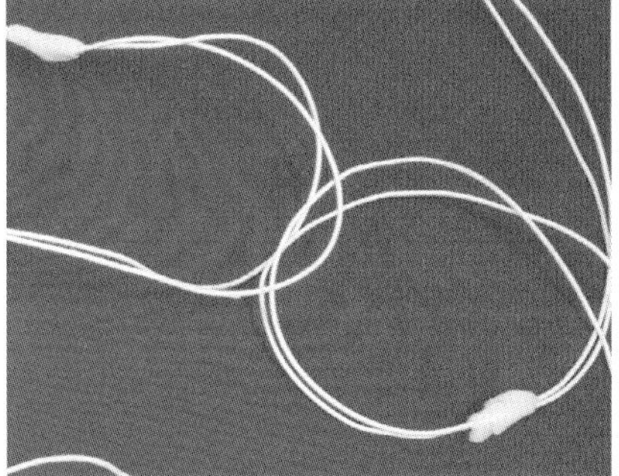

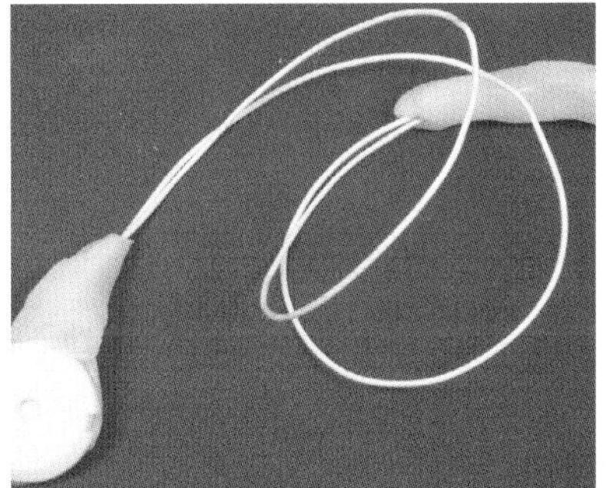

b. Heat some Polymorph and add small amounts intermittently along the pull string, securing the 2 wires together.

c. Using Polymorph, create a grip handle 4"–5" below the speaker. Next, create an encasement for the speaker, making sure not to cover the front of speaker (side with a circle). Take the solar crawler for a stroll and discover the radiant voice of the sun.

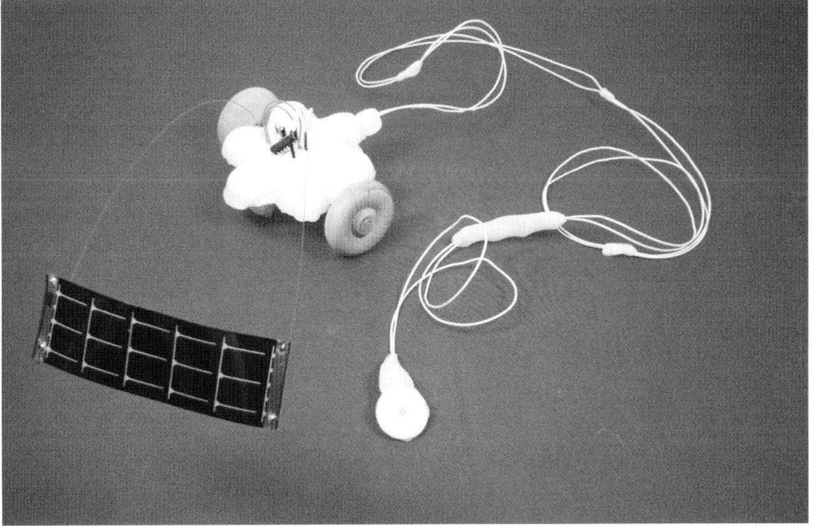

FINISH ⊠

SMART MOBILE

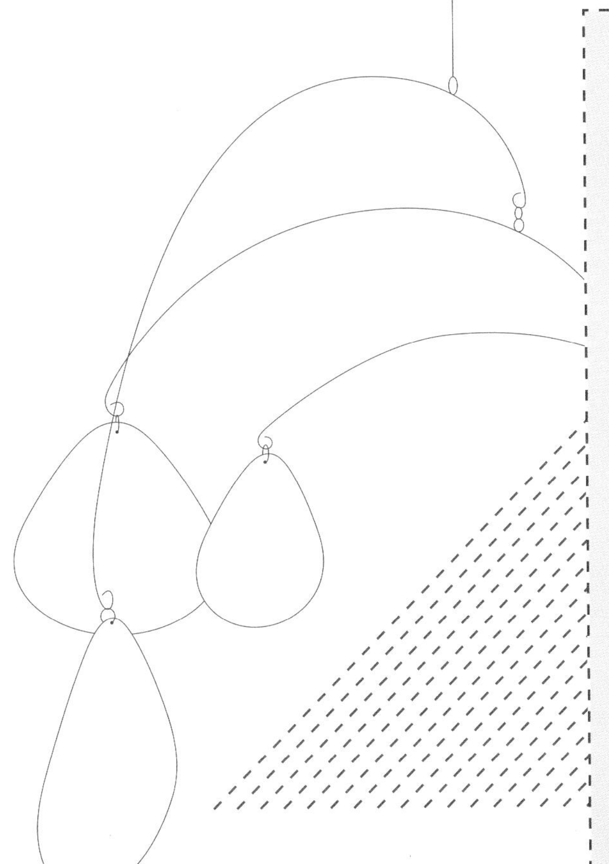

Playful and harmonious, mobiles embellish a space with a sense of lyrical serenity. With the help of smart wires, you can build and choreograph your own kinetic sculpture to dance exquisitely in space. The following tutorial will instruct you not only on how to construct and delicately balance the hanging elements of an intricate mobile, but, more interestingly, how to get your mobile to silently sway in a room without a window in sight.

HOW IT WORKS »

The Smart Mobile uses shape memory alloys (SMAs or "muscle wire") to trigger the mobile to sway and rock gently. SMAs typically exist in two states: relaxed and remembered. In the relaxed state, when no current is running through the wire to heat it, the wire is at normal length. In the remembered state, when heated by a current, the wire shortens. Once the wire cools again (no current), it reverts back to its relaxed state. The following tutorial uses a timing circuit to periodically heat and cool the wire, causing the mobile to twitch intermittently with small but perceptible movements.

COMPLEXITY: ADVANCED

WHAT YOU WILL LEARN

- Build a circuit on a perfboard
- Work with SMAs
- Make a mobile

RELATED TUTORIALS

- Art of Soldering (45-49)
- Building a Simple Circuit (27-33)
- Prototyping a Circuit Using a Perforated Board (33)

MATERIALS

- **18"×24" mat board**
- **8½"×11" sheets of cardstock (16)**
- **1" hook and loop**
- **Glue or spray mount**
- **Shape memory alloy (Flexinol), 4"** aka SMA or "muscle wire," part #141321, robotstore.com
- **Crimp beads (2)**
- **Solder**
- **Female sockets pin header (10-pin)** part #102201 jameco.com
- **Bag of 4.5mm metallic jump rings** as found in the jewelry supply section of any craft store
- **Malleable unshielded wire, 8¼', 10–12 AWG** for mobile
- **Embroidery thread**
- **Small screw eye hanger**
- **Bolt** must fit screw eye hanger

TEMPLATES

- Templates A–G
  Mobile petals (1 each)
- Template H
  Hanging frame (1)

TOOLS

- **Alligator clips**
- **Heavy duty wire cutters**
- **Sewing needle**
- **Soldering station**
- **Third hand** not necessary but highly recommended
- **Heat gun**
- **Scissors**
- **Utility knife**
- **Marker**
- **Needlenose pliers**
- **Wire cutters**
- **Wire strippers**
- **Hot glue gun**
- **Wire wrap tool**
- **Drill bit** (optional)

ELECTRONICS

- **LM555 timer IC** part #1301032 jameco.com
- **22µF capacitor**
- **1MΩ resistor**
- **1KΩ resistors (2)**
- **33Ω resistor**
- **LED**
- **Toggle switch**
- **9V battery connector**
- **9V battery**
- **2"×2" perfboard**
- **50' spool of 30 AWG wire wrap** recommended, but can be substituted for solid wire between 26–30 AWG
- **Variety of jumper wires or solid wire**
- **Solid wire, 3", 22 AWG**
- **16" heat-shrink tubing, ⅙" diameter**

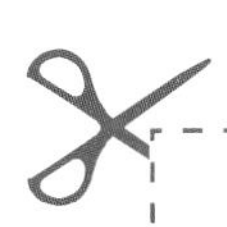

TEMPLATES

Available at fashioningtechnology.com/smartmobile.

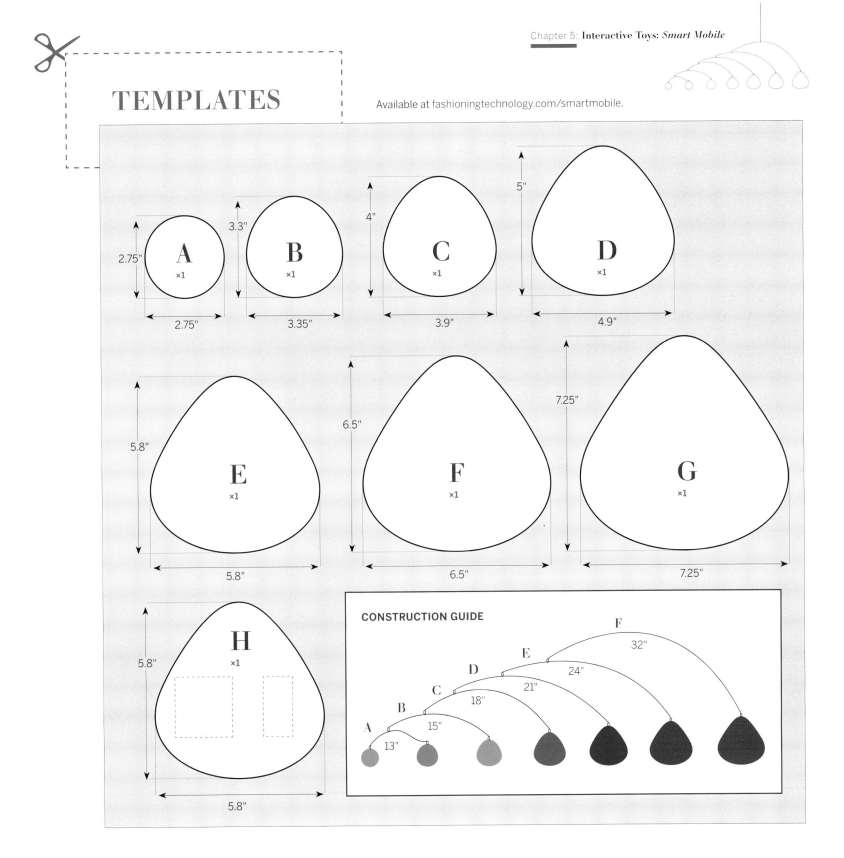

START »

1. MAKE THE FORM

a. Trace Templates A–H on the mat board. Using a utility knife, cut out all 8 pieces. Set Template H aside for now (you will using it later to construct the hanging structure for the mobile).

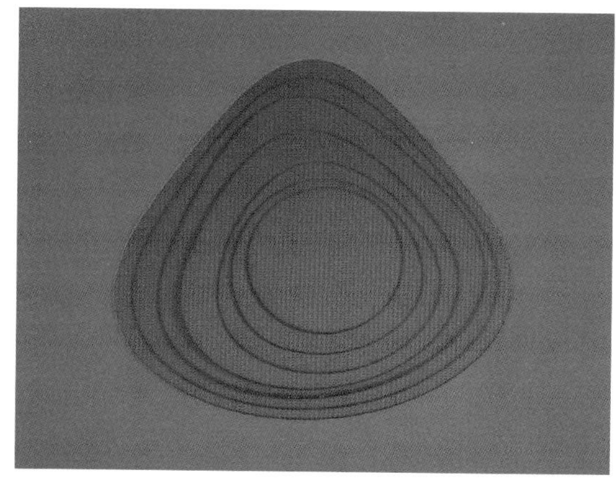

b. You can customize the mobile by covering the cut mat boards with decorative paper (or fabric). In the example provided, the cut templates form a gradient, from the lightest to darkest shade on one side and vice versa, from darkest to lightest on the opposite side. Print 2 of each template from A–G on the desired colored cardstock or trace them onto decorative fabric. Using a utility knife or scissors, cut out each template. Using the appropriate adhesive for the selected material, glue it directly on top of the cut mat board, making sure to cover both sides.

NOTE: To expedite the process, you can glue the colored pieces of cardstock onto the mat board first, with a different color on each side. Print 1 piece of colored cardstock with the template. Glue the cardstock with the template onto the mat board. Flip the board over, and glue another piece of colored cardstock on the opposite side. Cut. Alternately, you can use fabric. If you are using fabric, you will have to trace the template onto one side after you have glued the fabric onto the board.

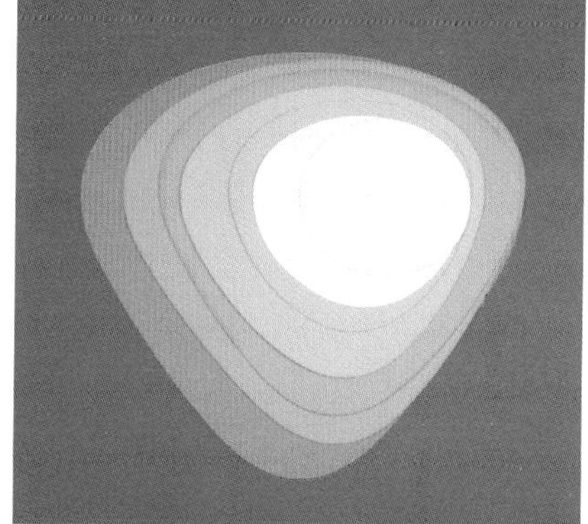

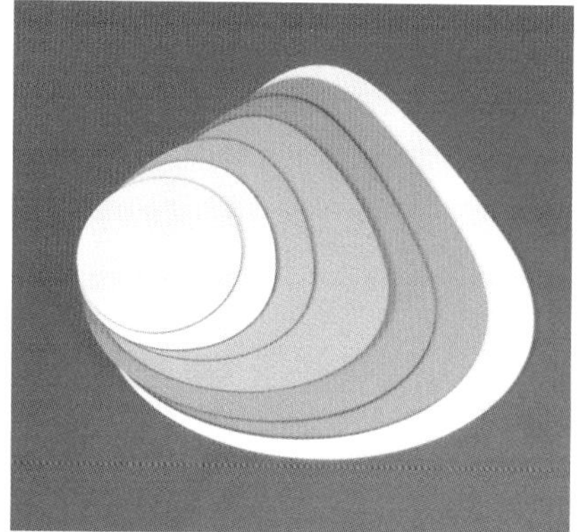

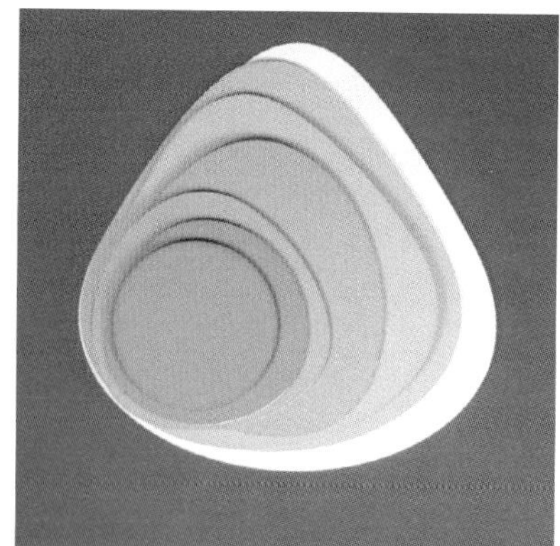

c. Using a sewing needle, pierce a hole approximately ¼" from the top of each pattern, *except* for Template H. Slip a metallic jump ring through each hole of Template H.

2. MAKE THE SWITCH

You will be wiring the switch onto two 36" pieces of 30 AWG wire, terminating at each end with a female pin. You will also be modifying the battery connector, extending the positive wire about 6", terminating with a male pin. Lastly, you will be making a 6" extension wire with a male pin on one end and a ½" piece of 22 AWG solid wire on the opposite end.

a. Using wire cutters, cut two 36" pieces of 30 AWG wire or wrap wire. Using wire strippers or the wire wrap tool (if you are using wire wrap wire), strip about ½" of insulation from both ends.

NOTE: The recommended length is based on an 8' standard ceiling height. If you have higher ceilings, you may need to make adjustments to the wire lengths accordingly.

? How do I use the wire wrap tool?

» Slip about 1" of wire into the slot in the middle of the wire wrap tool. Pull the wire down in between the metal teeth and then gently pull the wire towards you.

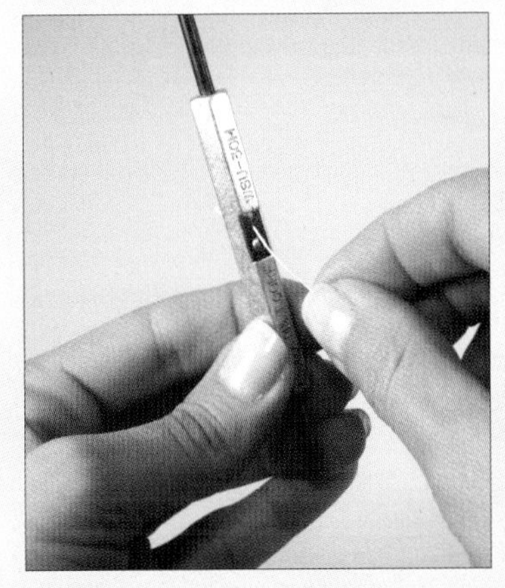

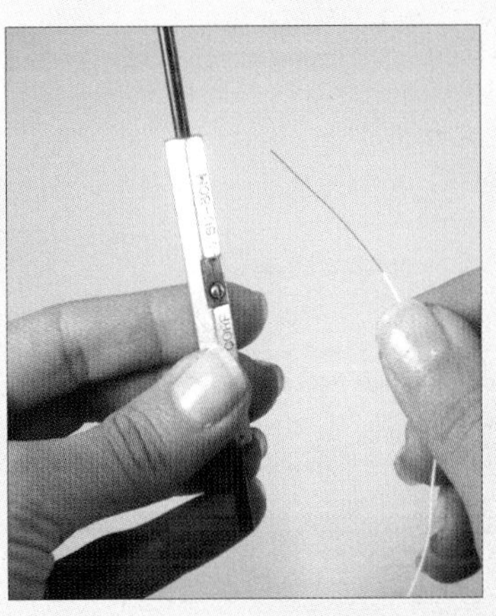

b. Using scissors, cut eight 1" pieces of heat-shrink tubing. Set aside. Wrap 1 stripped end of a wire onto a lead of the toggle switch and solder. Repeat for the second lead. Slip a 1" piece of heat-shrink tubing over each connection. Using a heat gun or blow dryer, shrink the tubing.

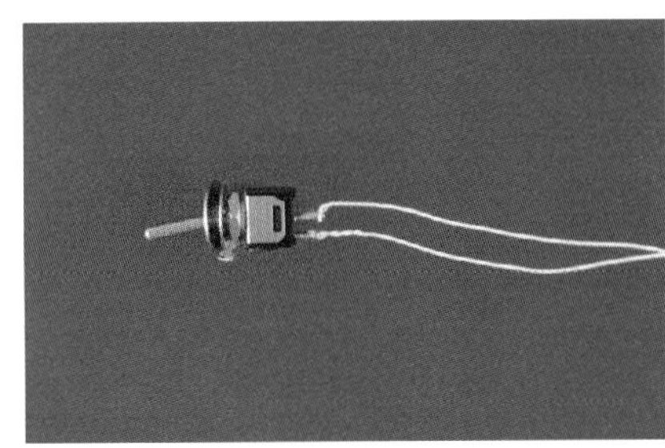

c. Using needlenose pliers, gently break off 1 pin from the 10-pin header. Repeat 3 more times, breaking off a total of 4 separate pins. Notice how each pin has a female and male counterpart. You will be soldering the switch wires around the male half of each pin.

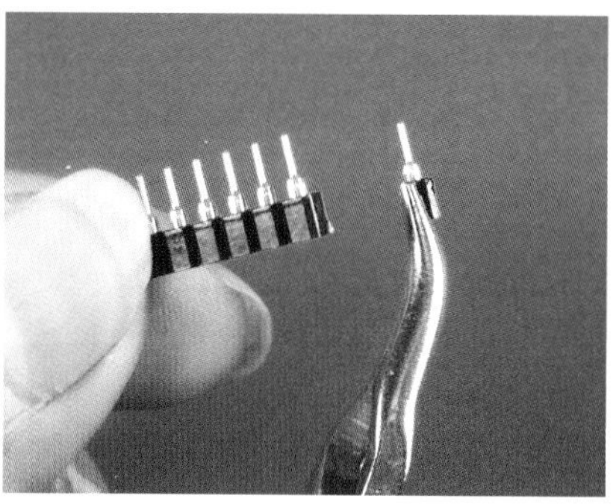

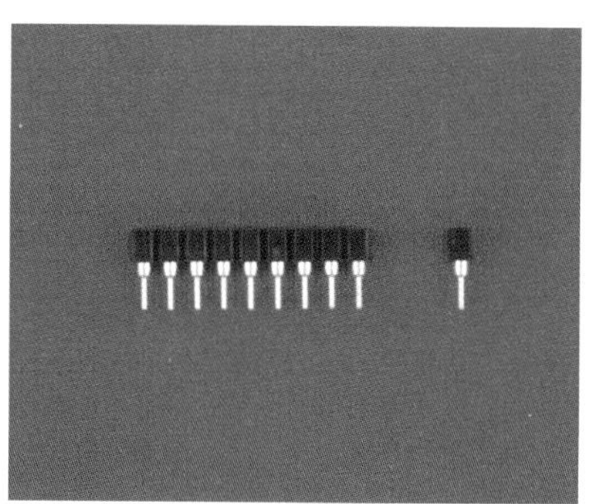

d. Clip 1 pin, with the male half facing up, onto the alligator clip of the third hand. Slip a 1" piece of heat-shrink tubing over one of the switch wires. Wrap the stripped end of the wire around the male end of the pin several times. Solder. Slip the heat-shrink tubing over the connection. Using a heat gun or blow dryer, shrink the tubing. Repeat, adding the second pin to the second switch wire. Both switch wires should now end with the female (not male) side of the pin.

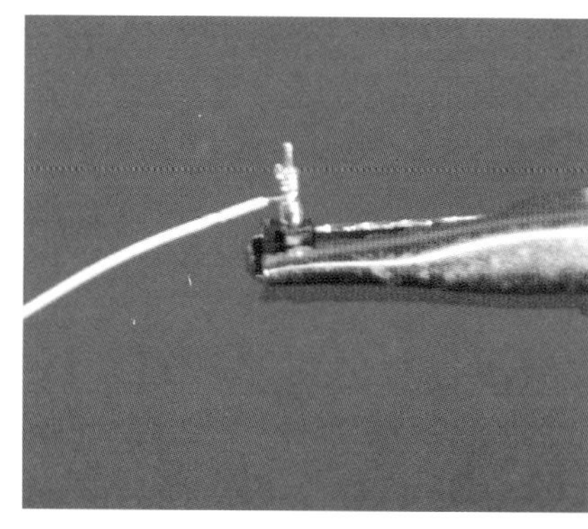

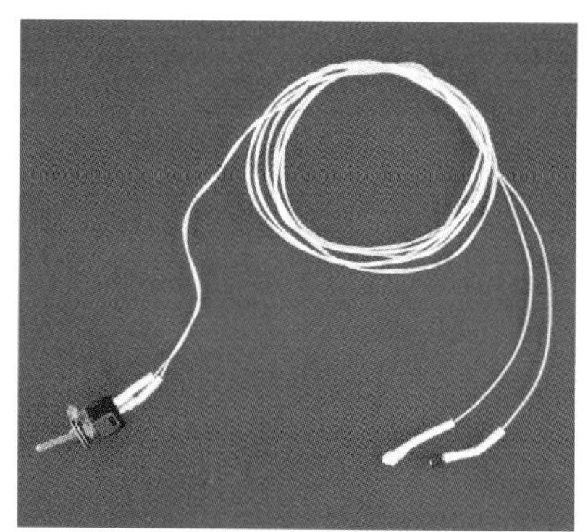

e. Cut 2 pieces of 6" 30 AWG wire and strip about ½" of insulation from both ends. Wrap a stripped end of a wire around the positive (red) lead of the 9V battery connector and solder. Slip the heat-shrink over the connection and shrink the tubing until snug. Clip the third pin, with the female half facing up, onto the alligator clip of the third hand. Slip a 1" piece of heat-shrink over the wire now connected to the positive battery lead. Place the wire into the female end of the pin and solder the wire securely in place. Slip the heat-shrink over the connection and shrink the tubing until snug. The positive battery connector wire should now end with the male side of the pin.

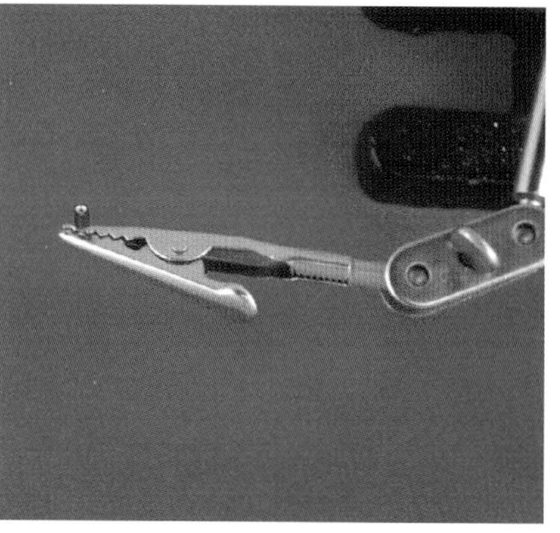

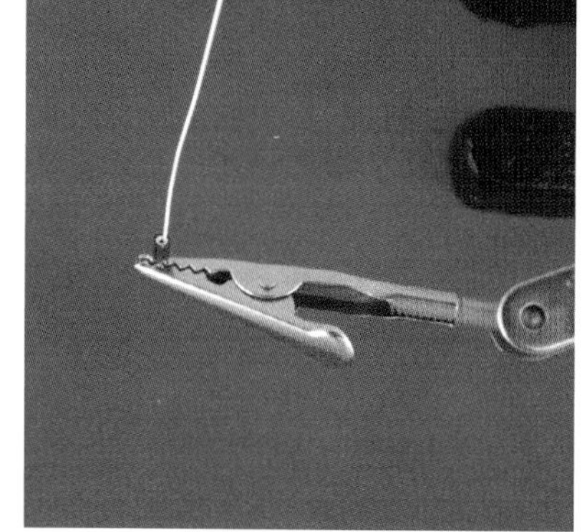

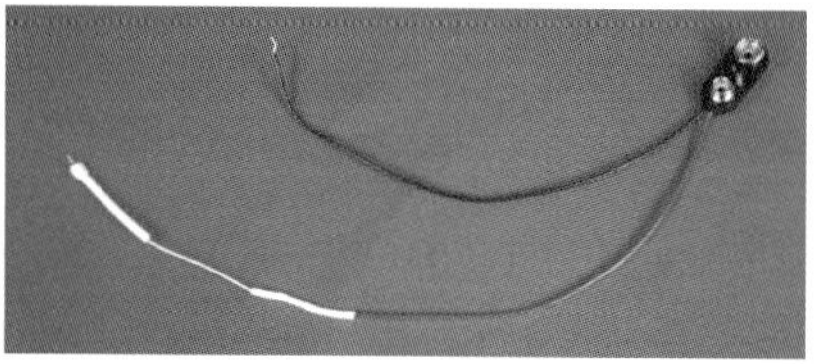

f. Now you will be making a 6" extension wire with a male pin on one end and a ¼" piece of 22 AWG solid wire on the opposite end. The 6" extension wire will be used to connect the circuit to the switch. The 22 AWG wire will make it easier for you to solder the extension wire to the perfboard. Take the remaining 6" piece of 30 AWG wire. Cut a 1" piece of solid 22 AWG wire and strip ¼" off each end. Wrap a stripped end of the 6" wire around one end of the solid wire. Solder. Slip a piece of heat-shrink over the connection and shrink. Slip another piece of heat-shrink over the opposite end of the wire. Place the last pin with the female half facing up in the third hand. Place the stripped end of the wire into the female end of the pin and solder. Cover the connection with heat-shrink tubing.

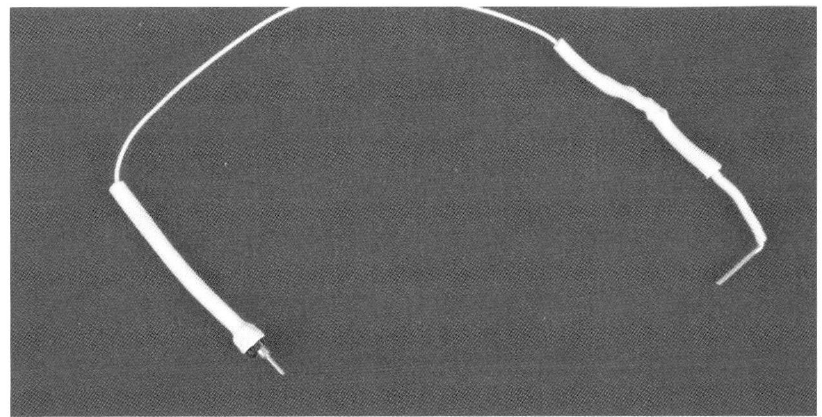

3. MAKE THE SMA MECHANISM

Heat from a soldering iron can damage shape memory alloys (SMAs, Flexinol in this case); therefore, alternative fastening methods must be employed to get a good electrical connection. Instead of soldering, use wire crimps to attach wire to both ends of the SMA. Similar to the switch, the wires will also terminate with a female pin on both ends.

a. Using wire cutters, cut two 4" pieces of 30 AWG wire and one 4" piece of Flexinol wire. Strip about 1" of insulation off both ends of each wire. Then, using scissors, cut four 1" pieces of heat-shrink tubing. Set aside.

b. Using needlenose pliers, bend the stripped end of a wire and 1 end of the Flexinol, and insert them both into a crimp bead. Compress the tube with pliers. Repeat for the opposite side, attaching the opposite end of the Flexinol to the second piece of wire using a crimp.

c. Using needlenose pliers, gently break off 2 more separate pins. Place a pin with the male half facing up in the third hand. Slip a piece of heat-shrink over a wire. Wrap the stripped end of one wire around the male end of the pin and solder. Cover the connection with heat-shrink tubing. Repeat for the second pin. You should now have a piece of Flexinol wire attached at each end to a piece of wire terminating with a female pin. Slip a piece of heat-shrink over the metallic crimps and shrink, covering all exposed metal.

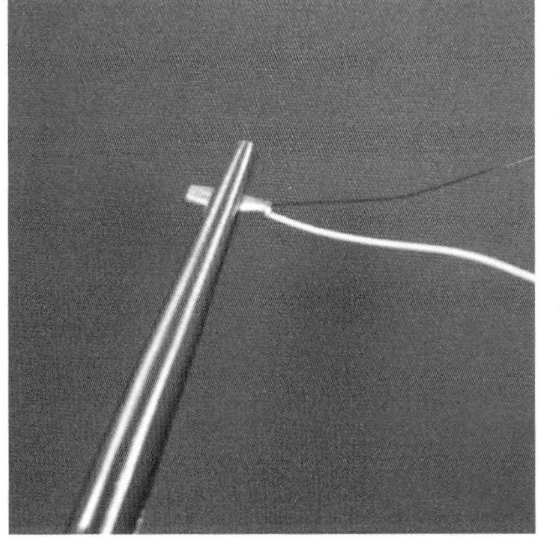

d. Now that you've made the Flexinol mechanism, you need to make the extension wires that will connect the muscle wire to the circuit board. The extension wires each will have a male pin at one end and a ½" piece of 22 AWG solid wire on the opposite end. The solid wire will make it easier for you to solder the thin extension wires to the circuit board. Using wire cutters, cut two 36" pieces of 30 AWG wire. Strip about 1" of insulation from both ends.

e. Cut two 1" pieces of 22 AWG solid wire. Strip about ¼" of insulation from both ends. Using scissors, cut four 1" pieces of heat-shrink tubing. Set aside.

f. Using needlenose pliers, gently break off 2 more pins. Place a pin with the female half facing up in the third hand tool. Slip a piece of heat-shrink over a 30 AWG wire. Place the stripped end of 1 wire into the female end of the pin and solder. Cover the connection with heat-shrink tubing. Repeat for the second pin, soldering the second 30 AWG wire.

g. Grab one of the 30 AWG wires and wrap the opposite end of the 30 AWG wire around 1 end of the solid 22 AWG wire. Solder. Slip a piece of heat-shrink over the connection and shrink. Repeat for the second wire. You should now have 2 separate 36" extension 30 AWG wires attached at 1 end to a piece of 22 AWG solid wire and on the opposite end to a male pin.

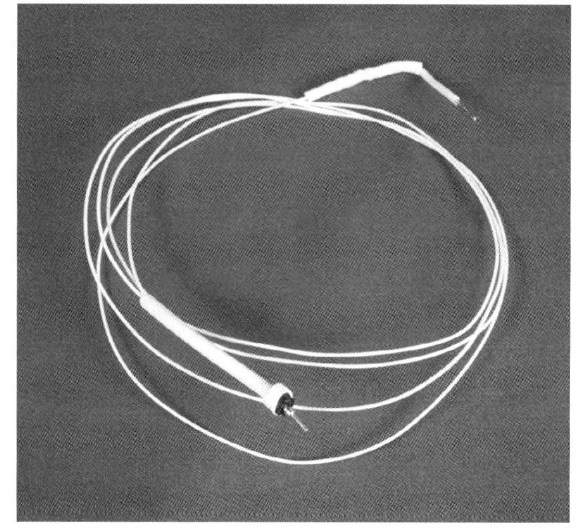

NOTE: Once again, the recommended length is based on the 8' standard ceiling height. If you have higher ceilings, make adjustments to the wire lengths accordingly.

4. BUILD THE CIRCUIT

If you have never built a circuit on a perforated board before, please review page 33 for more detailed information. This circuit diagram shows how the components of the circuit are connected to each other. It is not a direct translation of how the circuit will be laid out on the perforated board. Consider building the circuit on a breadboard first so you can better understand the circuit.

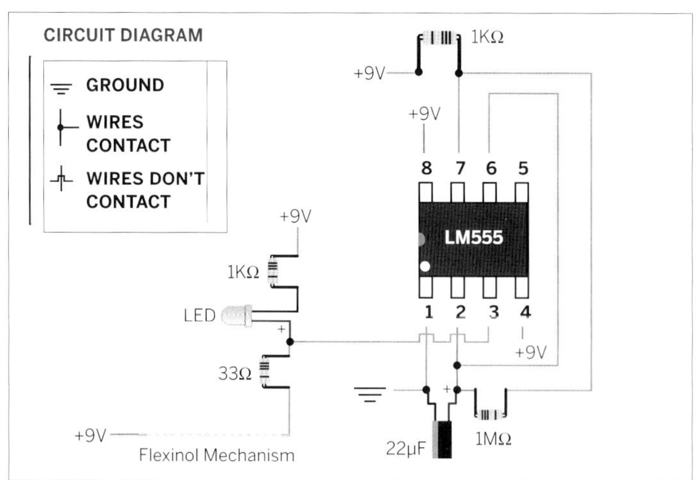

a. Grab the LM555 timer IC. Locate the notch and dot. The pins on the IC are numbered starting counterclockwise from the dot as in the circuit diagram. Using a perforated board designed with standard IC and component spacing, place the IC in the center of the perfboard, leaving a minimum of 3 rows in front of and behind the chip.

b. Designate a row for power and ground by placing a "+" and "−" near the assigned holes. In the shown example, both L-shaped rows on the left are the negative and the rows on the right are the positive.

c. First we are going to connect pin 2 to pin 6. Using jumper wire, jump pin 2 four rows to the right. Then jump pin 6 three rows to the right. Using jumper wire, connect the 2 rows that pins 2 and 6 have been jumped to (orange wire).

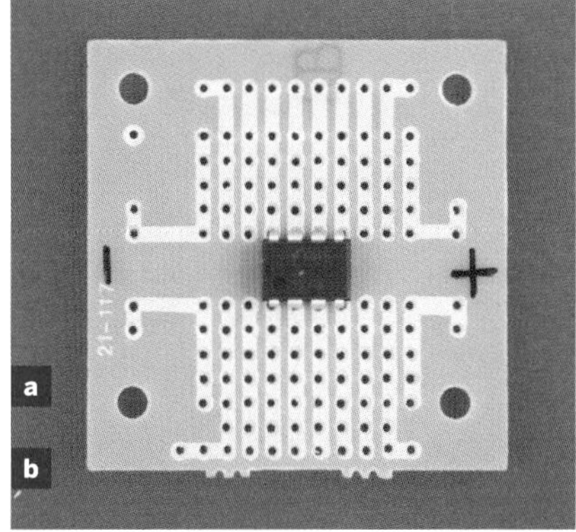

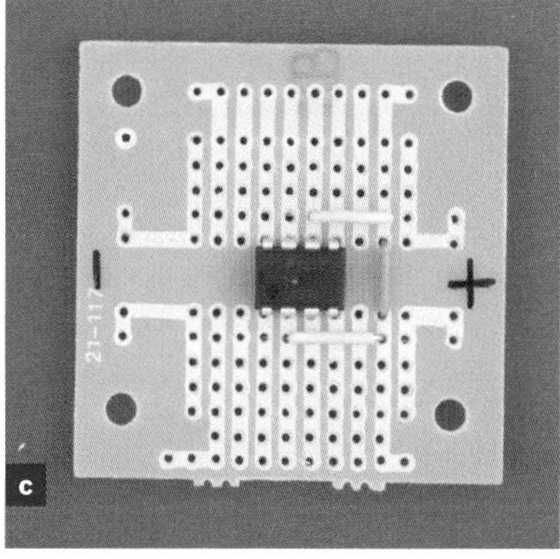

NOTE: The notch of the LM555 timer IC in the first photo is facing the "-" sign. The lower left pin is pin 1.

d. Next, wire the 1MΩ resistor between pins 2 and 7. To add a 1MΩ resistor, connect 1 lead of the resistor to pin 2, connecting the second lead over to the left 3 rows. To hold the resistor temporarily in place, bend the leads flush to the board. Using a jumper wire, jump pin 7 three rows to the left. Using another jumper wire, connect the second lead of the resistor to the row pin 7 has been jumped to.

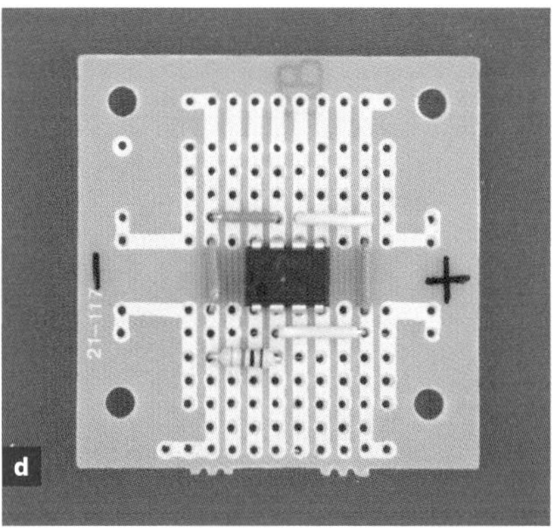

e. Connect the positive lead of the 22μF capacitor to pin 2 and the negative lead (the side marked with a gray bar) to pin 1. To hold the capacitor temporarily in place, bend the leads flush to the board.

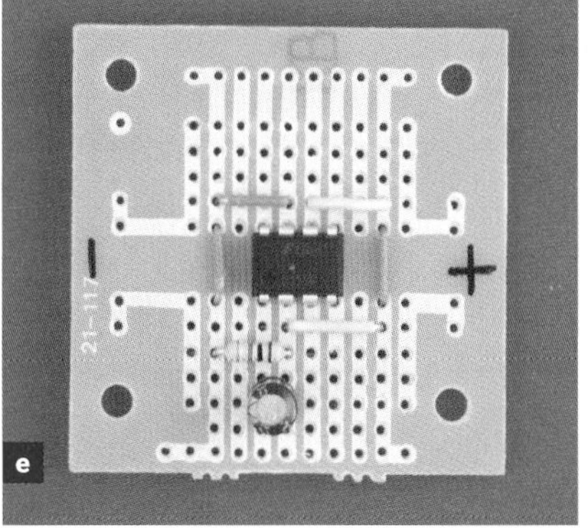

f. To connect pin 1 to ground, using a jumper wire, place the first lead of the wire in a hole under the negative lead of the capacitor (now connected to pin 1) and the second lead in any of the "–" holes.

g. To connect pin 8 to power (+9V), using a jumper wire, place the first lead of the jumper wire (blue) in the fourth hole above pin 8 and jump the row to any of the "+" holes.

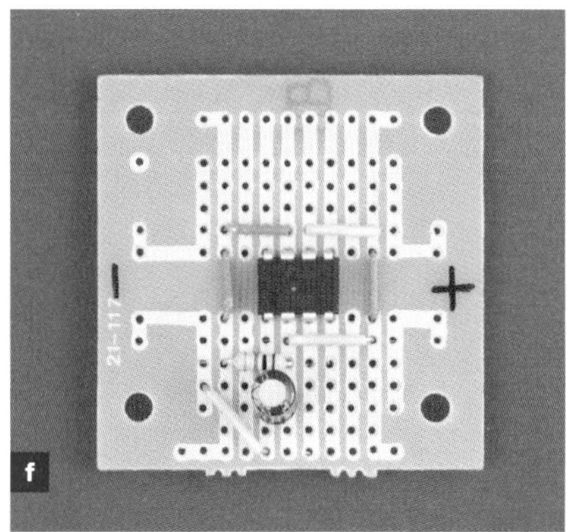
f

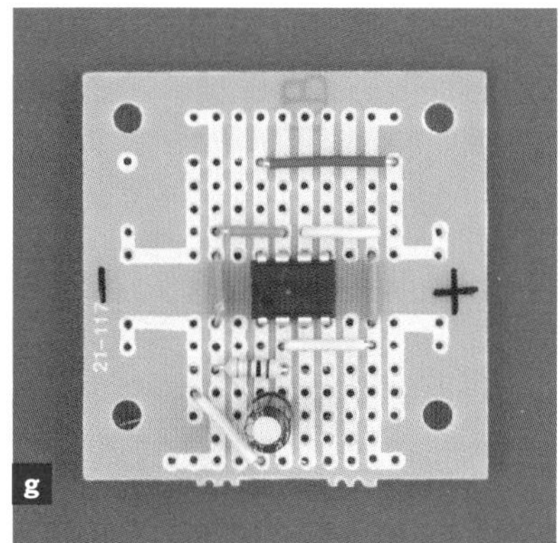
g

h. Connect the positive lead (longer lead) of the LED to pin 3 and the negative lead over 2 rows to the right to an empty row. Bend the leads flush to the board.

i. Connect 1 lead of a 1KΩ resistor to the negative LED lead, jumping the second lead to any of the "+" holes (to power).

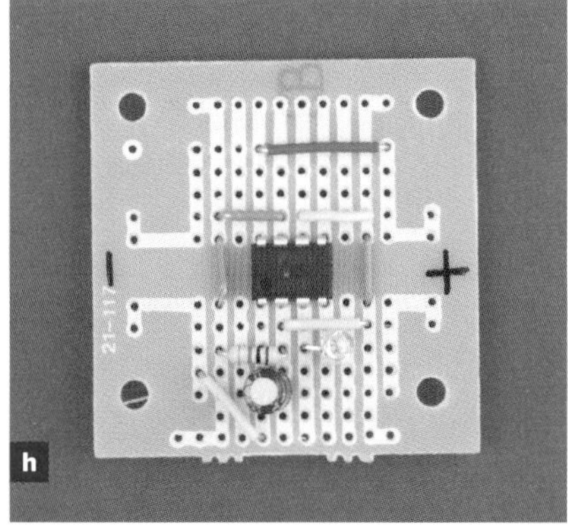
h

i

j. Connect 1 lead of the 33Ω resistor to pin 3 (right below the positive LED lead) and jump the second lead over to an empty row 3 rows to the left. Bend the leads flush to the board.

NOTE: This example uses 2 separate rows for power that must be connected. If your power rows are separated, use a jumper wire to connect the 2.

k. Connect 1 lead of the second 1KΩ resistor to pin 7 and jump the second lead to any of the "+" holes (power).

j

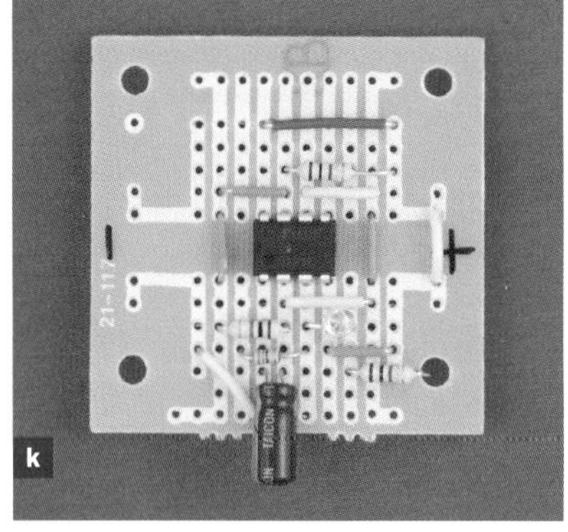
k

l. Grab one of the 36" extension wires for the Flexinol mechanism and connect the end with the solid 22 AWG wire to any of the "+" holes. Connect the second 36" extension wire to the row with the second lead of the 33Ω resistor (not to pin 3).

m. Connect the negative wire of the battery connector to any of the "–" holes (ground). Then take the 6" extension cable (made earlier), and connect the end with the solid 22 AWG wire to any of the "+" holes (power).

n. All the components in the circuit are now in place, and it's time to solder. Carefully turn the board around and solder all the leads of the components into place, making sure that the components don't move. Using wire cutters, trim all the excess once you have completed soldering.

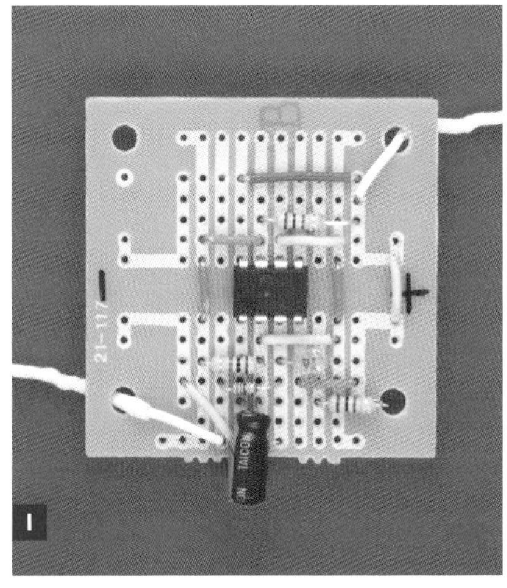
l

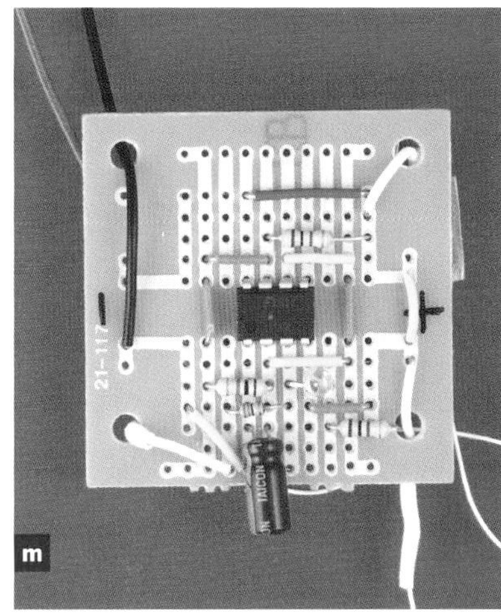
m

5. TEST THE CIRCUIT

Once you have soldered all the components to the board, it's a good idea to test the circuit. The LED on the board is used to indicate that the timing circuit is actually working.

a. Plug in one of the female ends of the toggle switch to the positive male end extending from the battery connector, and the other female end to the male end extending from the 6" extension cable (connected to a "+" hole on the board). Then plug the ends of the Flexinol mechanism to the 36" extension wires on the board.

b. Snap the battery into place. The LED light should light up about every 5–6 seconds intermittently. If the LED light doesn't turn on, flip the toggle switch. If you have switched the circuit on and it still doesn't work, unplug the battery and carefully review all your connections. Refer to the troubleshooting section on page 69.

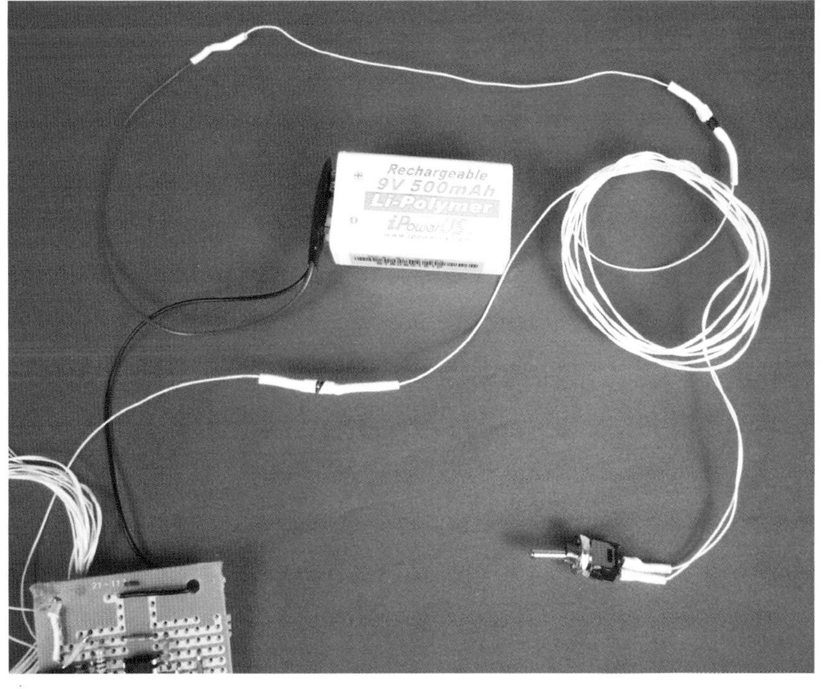

6. CONSTRUCT THE SCULPTURE

The key to successfully constructing any mobile is to start assembling the mobile from the bottom up. Each piece of the mobile must be carefully balanced with the others so that the mobile will hang correctly.

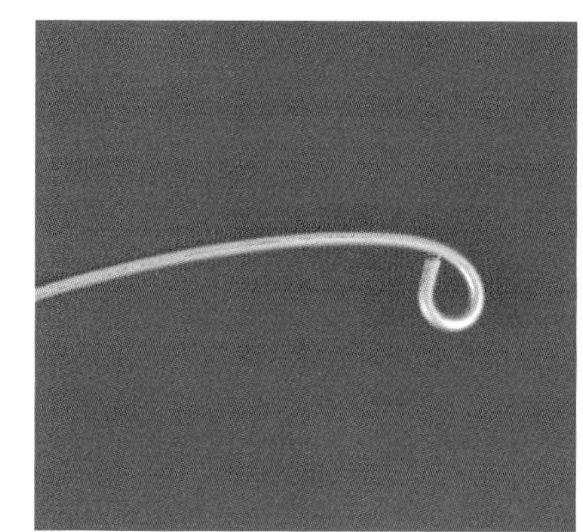

a. Using the construction guide on page 193, cut the 10–12 AWG mobile frame wires to the suggested lengths. You will need to use heavy duty wire cutters to be able to successfully cut through the thick wire. Then, using needlenose pliers, curl the ends of each frame wire in the direction suggested in the guide.

b. Gently shape Wire A according to the guide. Next, cut a 4" piece of thin solid 30 AWG wire. Using strippers, remove all of the insulation from the wire.

c. Secure a jump ring onto the upper curved portion of Wire A by wrapping the uninsulated wire around both the ring and frame wire. Add a touch of solder around the wires, securing the ring firmly to the frame wire.

d. Take the cutouts of Patterns A and B, and slip the jump rings around the curled ends of Wire A. Next, gently shape Wire B according to the guide. Secure the left curled loop of Wire B to the soldered loop of Wire A using a jump ring. The jump ring will allow the mobile pieces to twist and turn freely. Add the cutout of Template C to the right curled loop of Wire B.

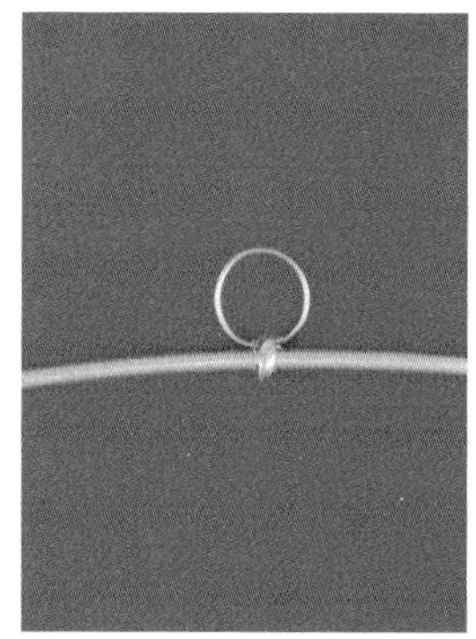

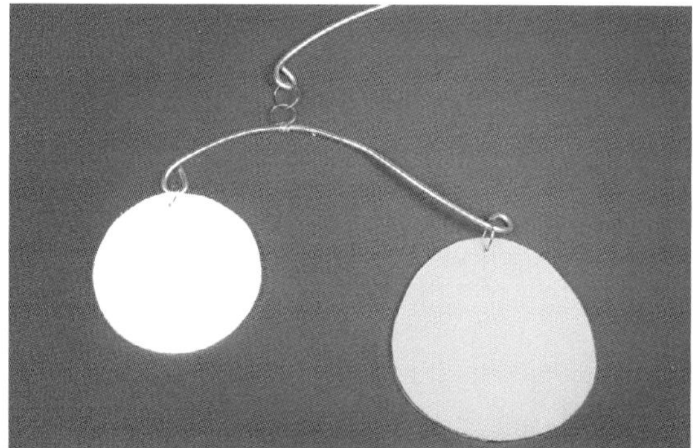

e. Now place an alligator clip in the center of Wire B. Hold the alligator clip from the opposite end, and keep adjusting the location where it is attached to Wire B until it hangs balanced. Using a marker, mark the location; you will be soldering the jump ring here.

f. Repeat the process for Wires C–F, carefully balancing each piece with the previous one.

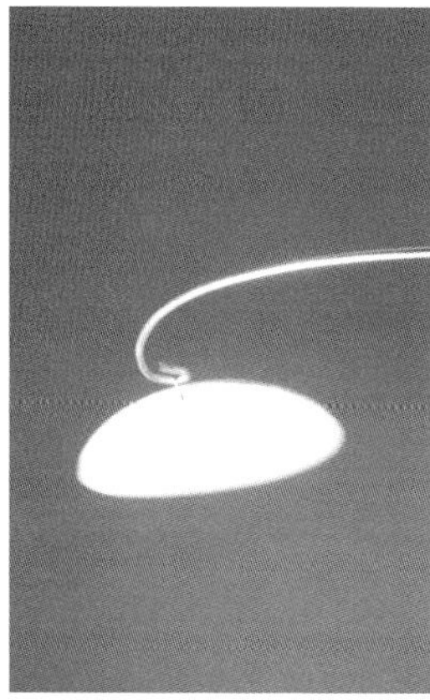

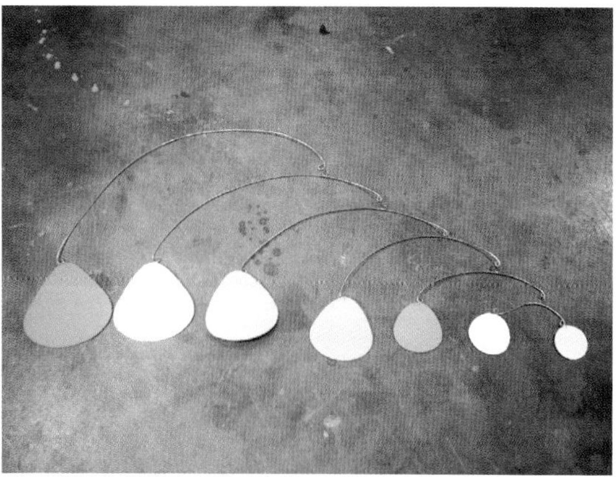

7. CREATE THE HANGING FRAME

All the electronics for the mobile will be placed on top of the hanging frame.

a. Take the cutout of Template H, and place the circuit board and the 9V battery in the suggested locations on the template. Secure the circuit board in place by adding a touch of hot glue to the ends of the board. Secure the battery in place by using a piece of hook and loop.

b. Using a drill bit or a sharp pointed object (such as the tip of a screw), pierce all the holes on Template H. Slip all the wires from the circuit board through the center hole.

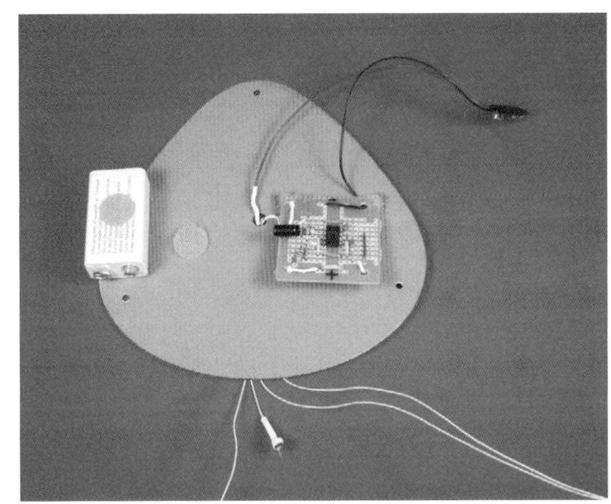

c. Cut three 12" lengths of embroidery thread and loop the threads through the outer pierced holes. These will function as the hanging mechanism for the entire mobile.

d. Slip the screw eye hanger from the bottom of the board through the center hole. Using a bolt, secure it in place. Then cut another piece of embroidery thread 3' in length. Knot it securely around the screw eye hanger (on the opposite side of the board).

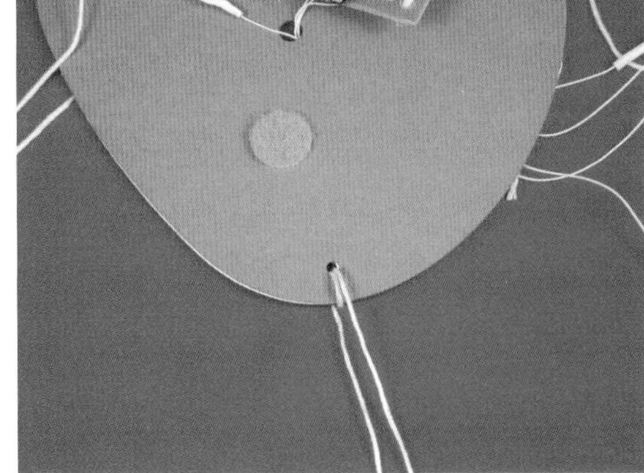

8. ASSEMBLE

The final assembly of the mobile requires that you temporarily hang the mobile low to the ground so that you can easily add the switch and Flexinol mechanism without too much difficulty. First determine how low you want to hang the mobile from the ceiling and adjust the length of the center embroidery thread accordingly.

a. Secure Wire F of the mobile assembly to the embroidery thread hanging from the center of the hanging frame. Cut the embroidery thread to its final desired length. Temporarily hang the entire assembly so that it is only 3' above the ground. If the switch and smart wire mechanism are still connected to the circuit board, carefully unplug them from the board.

b. Find a location where you want to position the switch on top of Wire F. Using a hot glue gun, secure the switch in place. Carefully wrap the switch wires around Wire F and then up the embroidery thread to the top of the circuit. Connect the switch back to its appropriate wires.

c. Wrap the Flexinol mechanism tightly around 2 joints of the mobile so that when the wires shorten, it causes the bottom wire to twist gently and the mobile to gently move. This part takes quite a bit of experimentation to find the perfect movement. Carefully wrap the 36" extension wires from the circuit down the center embroidery thread, connecting them to the Flexinol mechanism loosely. You don't want the connection wire to be too taut, or else it will limit the movement of the mobile.

d. Make sure that the battery is plugged in, and flip the switch on. Wait a few minutes until the mobile has settled. You should notice a periodic gentle sway, and occasional twisting, triggered by the shortening of the wire. If you haven't achieved the desired motion, try a different configuration of the Flexinol mechanism across 2 joints.

That's it! Hang your mobile and watch it twirl.

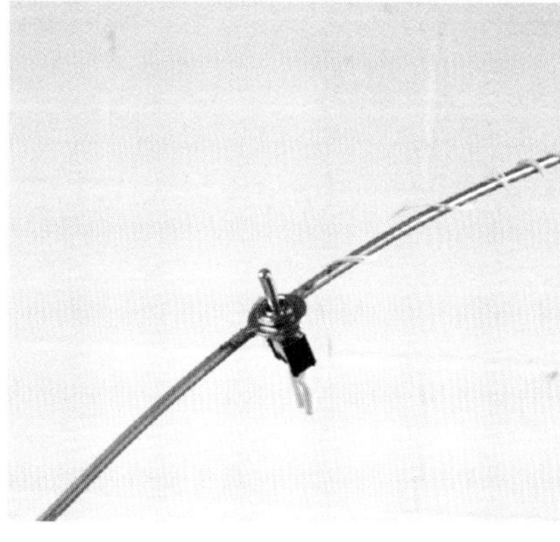

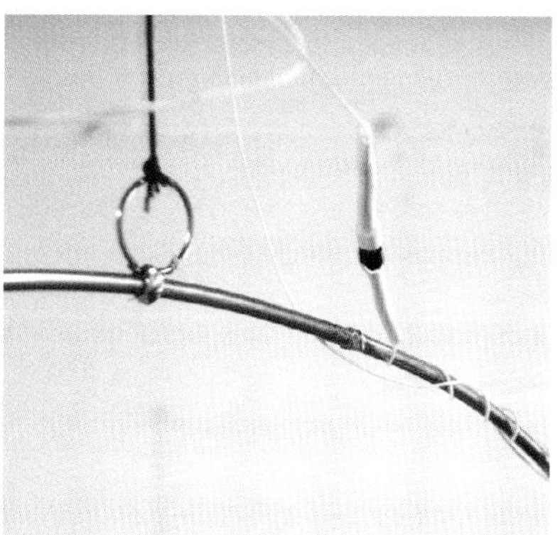

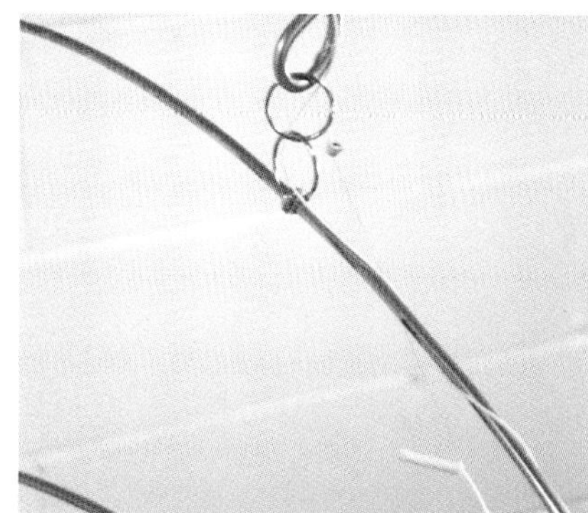

FINISH ⊠

WE'D LIKE TO HEAR FROM YOU

Please address comments and questions concerning this book to the publisher:

O'Reilly Media, Inc.
1005 Gravenstein Highway North
Sebastopol, CA 95472
(800) 998-9938 (in the United States or Canada)
(707) 829-0515 (international or local)
(707) 829-0104 (fax)

We have a website for this book, where we list errata, examples, and any additional information. You can access this page at: fashioningtechnology.com

To comment or ask technical questions about this book, send email to: bookquestions@oreilly.com

Maker Media is a division of O'Reilly Media devoted entirely to the growing community of resourceful people who believe that if you can imagine it, you can make it. Consisting of MAKE Magazine, CRAFT Magazine, Maker Faire, and the Hacks series of books, Maker Media encourages the Do-It-Yourself mentality by providing creative inspiration and instruction.

For more information about Maker Media, visit us online:
MAKE: makezine.com
CRAFT: craftzine.com
Maker Faire: makerfaire.com
Hacks: hackszine.com

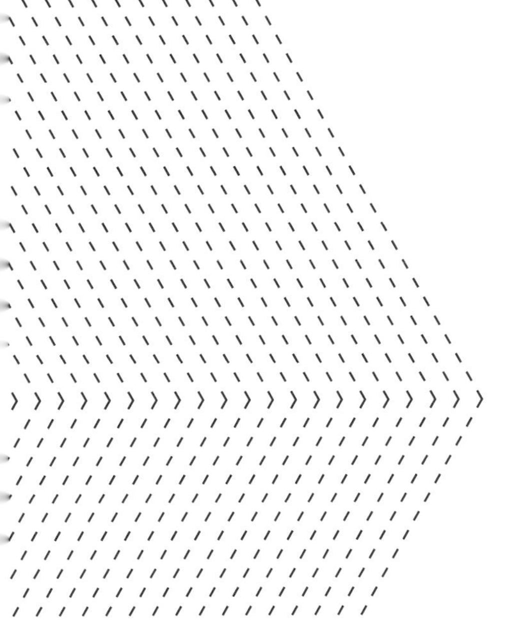

RESOURCES

RESOURCES: LED RESISTOR CALCULATORS

There are several resources available online that calculate the appropriate resistor value you should use in your LED circuits. Here are three good ones. Refer to page 21 for more information on resistor calculators.

ledcalc.com
quickar.com/noqbestledcalc.htm
myresistor.com

RESOURCES: SUPPLIERS

» BATTERIES AND SOLAR CELLS

All-Battery
all-battery.com
Retailer of practically every type and size of battery, as well as solar panels.

FlexSolarCells
flexsolarcells.com
Retailer of flexible solar cells and panels.

Plastecs Company
plastecs.com
Retailer of inexpensive solar cells, panels, and kits, as well as fiber optics and fiber optic kits.

PowerStream
powerstream.com
Retailer of batteries in unique packaging.

Solar World
solarworld.com
Retailer of a variety of solar panels, products, and kits.

Solarbotics
solarbotics.com
Retailer of robot kits, solar-powered robots, solar cells, and electronic components.

Sundance Solar
sundancesolar.com
Retailer of a variety of solar panels, products, and kits.

» CONDUCTIVE HOOK AND LOOP

Block EMF
blockemf.com
Retailer of conductive fabrics, fabric tapes (nickel/copper fabric tape), and conductive hook and loop.

Fastech of Jacksonville, Inc.
hookandloop.com
Retailer of various types of hook and loop, including conductive.

Less EMF Inc.
lessemf.com
Retailer of conductive fabrics, tapes, thread, inks, paints, hook and loop, and epoxies.

» CONDUCTIVE THREAD AND FABRIC

Bekaert
bekaert.com
Manufacturer specializing in electrically conductive technical textiles and yarns.

Fine Silver Products
fine-silver-productsnet.com
Retailer of Shieldex conductive fabrics, yarns, and zipper tapes.

Lamé Lifesaver
members.shaw.ca/ubik/thread/index.html
Supplier of excellent conductive thread with low resistivity.

Less EMF Inc.
lessemf.com
Retailer of conductive fabrics, tapes, threads, inks, paints, velcro, and epoxies. Great source to begin experimenting with various conductive materials.

Shieldex Trading
shieldextrading.net
Manufacturer specializing in metalized (plated with metal) conductive fabrics, yarns, and zipper tapes.

SparkFun Electronics
sparkfun.com
Great resource for kits, sensors, electronic components, and conductive thread.

X-Static
x-staticfiber.com
Manufacturer specializing in fiber that is 99.9% pure silver.

» ELECTROLUMINESCENT WIRE, FILM, AND TAPE

CooLight
coolight.com
Retailer of EL wire by the foot, LEDs, sequencers, and inverters. Kits also available.

Elwire
elwire.com
Great site for learning all about EL wire. Retailer of spools of EL wire.

VibeLights
vibelights.com
Retailier of EL tape and EL wire by the foot. Wholesale pricing available.

»ELECTRONICS

All Electronics
allelectronics.com
Retailer of electronics and electro-mechanical parts. Good pricing.

Digi-Key
digikey.com
All-encompassing electronics retailer. Fast shipping.

Electronic Goldmine
goldmine-elec.com
Excellent resource for surplus electronics and unique LED packages and sensors. Great resource for cellphone batteries.

Jameco Electronics
jameco.com
Great resource for electronic components, prototyping supplies, and tools. Order their catalog, as the search engine is difficult to use.

Mouser Electronics
mouser.com
Electronics retailer with virtually everything you could need.

Sparkfun Electronics
sparkfun.com
Electronics supplier catering to developers and prototypers. Great resource for kits, sensors, components, and conductive thread.

» FIBER OPTICS

The Fiber Optic Store
thefiberopticstore.com
Supplier of clear fiber optic filaments in spools.

Oakridge Hobbies
oakridgehobbies.com
Excellent resource for clear and fluorescent fiber optics, acrylic sheets, and rods.

Plastecs Company
plastecs.com
Retailer of inexpensive fiber optics and fiber optic kits.

»INDUSTRIAL FELT, FABRICS, AND MATERIALS

Aetna Felt Corporation
aetnafelt.com
Manufacturer and retailer of 100% wool designer industrial felts available in a variety of colors and sold by the yard.

Dazian Fabrics
dazian.com
Retailer specializing in flame-retardant, wide-width fabrics.

The Felt People
thefeltpeople.com
Supplier of decorative and industrial felts and fabrics.

Orchard Supply Hardware
osh.com
Retailer with a good selection of industrial materials, including plexiglass, felt, and cotter pins.

Sutherland Felt Company
sutherlandfelt.com
Supplier of natural industrial wool felts available in a various widths, densities, and thickness, sold by the yard.

»LEDS

HB
hebeiltd.com
Great, inexpensive resource from China for LEDs. Slow shipping. Carries high-flux Piranha LEDs.

Lumex
lumex.com
High-quality LED supplier with unique LED packages.

Super Bright LEDs, Inc.
superbrightleds.com
Great resource for all types of LEDs. Carries high-flux Piranha LEDs.

»MAGNETS, MAGNETIC PAINT, AND MAGNETIC LIQUID

Amazing Magnets
amazingmagnets.com
Excellent resource for rare earth magnets, available in a variety of shapes, from rods to spheres.

The Container Store
containerstore.com
Great resource for flexible magnetic strips and mighty magnets.

MagnaMagic
magnamagic.com
Manufacturer of magnetic and chalkboard paints.

MUTR
mutr.co.uk
Excellent educational resource for electronic components and kits, smart materials, and more.

» PHOSPHORESCENT MATERIALS

Brightec
brightec.com
Manufacturer of high quality glow-in-the-dark inkjet paper.

McLogan Supply Company
mclogan.com
From screens and squeegees to specialty phosphorescent inks, everything you need to set up a screen-printing studio at home.

MUTR
mutr.co.uk
Great place for small quantities of thermo- and photochromic powders and inks.

Risk Reactor
riskreactor.com
Retailer of UV and phosphorescent pigments and dyes.

»POLYMORPH PLASTIC (aka ShapeLock or Friendly Plastic)

The Compleat Sculptor
sculpt.com
Retailer of Polymorph "friendly" plastic.

MUTR
mutr.co.uk
Excellent educational resource for electronic components and kits, smart materials, and more.

ShapeLock
shapelock.com
Manufacturer and retailer of Polymorph plastic.

Sunshine Discount Crafts
sunshinecrafts.com
Retailer of "friendly plastic" sticks, available in a variety of colors.

» SCREEN-PRINTING EQUIPMENT

McLogan Supply Company
mclogan.com
From screens and squeegees to speciality phosphorescent inks, everything you need to set up a screen-printing studio at home.

» SHAPE MEMORY ALLOYS (SMA)

Dynalloy, Inc.
dynalloy.com
Manufacturer of Flexinol and Muscle Wire SMAs. Have their own kits and sell wire by the foot.

Edmund Scientific
scientificsonline.com
Supplier of teaching resources from solar kits to gears, magnets, and Nitinol SMA.

Jameco Robot Store
robotstore.com
Offshoot of Jameco with good supply of SMA kits and wires by the meter.

» STORAGE/ ORGANIZATION

The Container Store
containerstore.com
Excellent for storage and organization containers for crafting and electronic components.

» THERMOCHROMIC AND PHOTOCHROMIC MATERIALS

Color Change Corporation
colorchange.com
Specializing in thermochromic dyes and photochromic inks. Can order small-quantity sample packs.

Chromatic Technologies
ctiinks.com
Manufacturer of thermochromic, photochromic, and photoluminescent inks.

Educational Innovations
teachersource.com
Good resource for photochromic beads and magnetic liquid.

Embroider This!
embroiderthis.com
Supplier of photochromic embroidery thread.

LDP LLC
maxmax.com
Retailer of photochromic inks and powders, photoluminescent inks, and magnetic inks.

MUTR
mutr.co.uk
Great place to purchase small quantities of thermo- and photochromic powders and inks.

SolarActive International
solaractiveintl.com
Supplier of all types of photochromic materials, including inks, threads, beads, nail polishes, and fabrics.

Textura Trading Company
texturatrading.com
Great supply of unique industrial yarns, including UV, reflective, and blends with steel.

Union Ink
unionink.com
Manufacturer of thermochromic, photochromic, and photoluminescent inks. Great supply of specialty inks, such as Chalkboard and Crayon Clear.

» TOOLS

Jameco Electronics
jameco.com
Prime resource for electronic components, prototyping supplies, and tools. Order their catalog, as the search engine is difficult to use.

RESOURCES: BOOKS

Electronics Fundamentals: Circuits, Devices, and Applications
By Thomas L. Floyd
Prentice Hall

Extreme Textiles: Design for High Performance
By Matilda McQuaid
Princeton Architectural Press

Fashioning the Future: Tomorrow's Wardrobe
By Suzanne Lee
Thames & Hudson

Material World: Innovative Structures and Finishes for Interiors
By Edwin van Onna
Birkhäuser Press

Making Things Talk: Practical Methods for Connecting Physical Objects
By Tom Igoe
O'Reilly Media

Muscle Wire Project Book
By Roger G. Gilbertson
Mondo-Tronics, Inc.

Petit Pattern Books: Dots and Stripes, Check and Knit, Flowers and Leaves, Japanese Style
Bnn Pattern Book Series
Ram Distribution

Physical Computing: Sensing and Controlling the Physical World with Computers
By Tom Igoe and Dan O'Sullivan
Course Technology PTR

Smart Materials in Architecture, Interior Architecture and Design
By Axel Ritter
Birkhäuser Press

Techno Fashion
By Bradley Quinn
Berg Publishers

Techno Textiles 2: Revolutionary Fabrics for Fashion and Design
By Sarah E. Braddock Clarke and Marie O'Mahony
Thames & Hudson

Transmaterial 2: A Catalog of Materials That Redefine Our Physical Environment
By Blaine Brownell
Princeton Architectural Press

Vintage Fabric from the States
By Various
P.I.E. Books (available through Amazon UK)

AUTHOR BIOGRAPHY

Syuzi Pakhchyan is a user experience designer and tinkerer working and residing in Los Angeles, Calif. Her work explores the intersection of culture and technology through the experimentation, investigation, and design of interactive technological systems for a range of cultural contexts.

She received her BFA from UC Berkeley and her MFA in Media Design from the Art Center College of Design. Her work in fashion and technology has been exhibited at Eyebeam, an art and technology gallery in New York, as well as at the Fashion Future Event, Maker Faire, and the O'Reilly Emerging Technologies Conference.

Her designs explore and encourage ludic activities that celebrate the quirky and the speculative, and reflect on personal experiences as well as cultural narratives.

Using design as provocation, Syuzi has created a diverse body of open-ended works that deconstruct traditional inanimate/animate relationships and reconfigure them into unexpected artifacts and experiences. From *Love Objects* — lamps that you blow on to light and caress to increase brightness — to the creation of sewn, decorative electronic circuitry, her work addresses our (im)personal and (un)intimate relationships with the material world, and attempts to reinfuse the seemingly objective with subjectivity.

Currently, Syuzi is working as a freelance user experience and interaction design consultant and teaches a robotics class to children.

CONTRIBUTORS

Ralf Schreiber lives in Cologne, Germany, and works with audio installations, robotics, chaotic processes, auto active systems, and silence. Ralf finished his MA studies at the Münster College of Art and his postgraduate studies at the Academy of Media Arts Cologne.

Ralf has exhibited in several galleries and international festivals of media arts (recent selection): The Eye of Sound, Museum Schoss Moyland, 2007; Happy New Ears/Sonokids, Belgium, 2007; Interferenze 4, San Martino Valle Caudina, Italy, 2006; Wake Up, Rauma Biennale Balticum, Rauma, Finland, 2006; STRP Festival Eindhoven, Netherlands, 2006. ralfschreiber.com

Ralf is the genius behind the the solar sound module circuit used to create both Aerial the Birdie Brooch and the Solar Crawler.

Jed Berk is an artist, inventor, and entrepreneur whose work lies at the intersection of art, technology, and entertainment. Jed explores ideas related to the transient point in nature where evolution might occur. He employs emerging technologies to create biotopes of semi-domesticated, biologically inspired sculptures that live in network-based ecosystems.

Born in New York, Jed now lives in Los Angeles, Calif. He received a BFA from the Rhode Island School of Design and an MFA from Art Center College of Design. He has shown in the U.S. and internationally. Jed's work has reached a broad audience with the ALAVs international television debut on the Discovery Network. degree119.com

Jed helped inspire the development of the Solar Crawler.

Incorporating illustrations created by hand, **Sara Schmidt**'s work transforms places into enchanting, whimsical atmospheres. At this very moment, she is likely doodling up a storm to create new patterns for use in women's and children's apparel, while Cooper, her lovable orange tabby, attempts to interrupt. Sara has an MFA from Art Center College of Design and lives and works in Portland, Ore. Learn more about Sara at salvage.la and playmakecreate.com.

Sara created the delightful pattern for the Luminescent Tea Table.

INDEX

M

N

O

P

R

NOTES: